Environmental Pollutants in the Mediterranean Sea: Recent Trends and Remediation Approaches

Edited By

Tamer El-Sayed Ali

Department of Oceanography
Faculty of Science
Alexandria University
Alexandria, Egypt

Environmental Pollutants in the Mediterranean Sea: Recent Trends and Remediation Approaches

Editor: Tamer El-Sayed Ali

ISBN (Online): 978-981-5179-06-4

ISBN (Print): 978-981-5179-07-1

ISBN (Paperback): 978-981-5179-08-8

need for a court order if at any point you breach any terms of this License Agreement. In no event will any delay or failure by Bentham Science Publishers in enforcing your compliance with this License Agreement constitute a waiver of any of its rights.

3. You acknowledge that you have read this License Agreement, and agree to be bound by its terms and conditions. To the extent that any other terms and conditions presented on any website of Bentham Science Publishers conflict with, or are inconsistent with, the terms and conditions set out in this License Agreement, you acknowledge that the terms and conditions set out in this License Agreement shall prevail.

Bentham Science Publishers Pte. Ltd.
80 Robinson Road #02-00
Singapore 068898
Singapore
Email: subscriptions@benthamscience.net

BENTHAM SCIENCE

CONTENTS

PREFACE

Pollution of the aquatic environment is a real threat across the globe. It is becoming a topic of intense study for researchers. It has been updated and almost completely revised. The Mediterranean Sea has been recognized as a target hotspot of the world as the pollutants' concentration in this region is greater than other vast oceans.

The book summarizes the basic pollutants in the Mediterranean Sea, focusing on microplastics, rare earth elements and biotoxins. The impacts of pollutants on the aquatic environment and health have received greater attention in this edition, and there are more case descriptions that cover some countries in the Mediterranean region. This edition has a simple emphasis on pollution problems in the Mediterranean region. The present book is an attempt to put together, to a certain level, the scientific publications on the referred topic for the benefit of the target audience. In addition, more references are included at the end of the chapters, including many websites.

A framework: Chapters 1 through 5 provide basic information on pollutants in the Mediterranean Sea with a special focus on some countries in the region and how humans are affecting the environment. They also discuss important concepts of the pollutant's hazards. Chapter 1 demonstrates the impact of Biotoxins in Lebanon as a case study. Chapter 2 discusses the bio-monitoring of effluents. Chapter 3 introduces basic concepts in microplastic pollution and its effect on fish while Chapter 4 explains the impact of microplastics on the Moroccan Mediterranean. Chapter 5 delves into the ecological impacts of rare earth elements influx into the Mediterranean Sea.

I hope readers who are interested in environmental pollution in general, or involved in aquatic science in particular, can find this book useful and objective achievable. Finally, I would very much welcome any feedback from readers.

Tamer El-Sayed Ali
Department of Oceanography
Faculty of Science
Alexandria University
Alexandria, Egypt

List of Contributors

Abed El Rahman Hassoun	National Council for Scientific Research, National Center for Marine Sciences, Batroun, Lebanon
Ahmed Mandour	Department of Oceanography, Faculty of Science, Alexandria University, Baghdad St., Moharem Bek, 21511, Alexandria, Egypt
Bilal Mghili	LESCB, URL-CNRST N° 18, Abdelmalek Essaadi University, Faculty of Sciences, Tetouan, Morocco
Céline Mahfouz	National Council for Scientific Research, National Center for Marine Sciences, Batroun, Lebanon
Ivana Ujević	Laboratory of Plankton and Shellfish Toxicity, Institute of Oceanography and Fisheries, Šetalište Ivana Meštrovića 63, Split, Croatia
Milad Fakhri	National Council for Scientific Research, National Center for Marine Sciences, Batroun, Lebanon
Mohamed Analla	LESCB, URL-CNRST N° 18, Abdelmalek Essaadi University, Faculty of Sciences, Tetouan, Morocco
Monia El Bour	INSTM, Rue du 2 mars 1934, 2025, Salammbo, Tunisia
Mustapha Aksissou	LESCB, URL-CNRST N° 18, Abdelmalek Essaadi University, Faculty of Sciences, Tetouan, Morocco
Nikša Nazlić	Laboratory of Plankton and Shellfish Toxicity, Institute of Oceanography and Fisheries, Šetalište Ivana Meštrovića 63, Split, Croatia
Romana Roje-Busatto	Laboratory of Plankton and Shellfish Toxicity, Institute of Oceanography and Fisheries, Šetalište Ivana Meštrovića 63, Split, Croatia
Sahar Karray	INSTM, Rue du 2 mars 1934, 2025, Salammbo, Tunisia
Sharif Jemaa	National Council for Scientific Research, National Center for Marine Sciences, Batroun, Lebanon
Tamer El-Sayed Ali	Department of Oceanography, Faculty of Science, Alexandria University, Alexandria, Egypt

CHAPTER 1

Biotoxins in the Mediterranean Sea: Lebanon as a Case Study

Abed El Rahman Hassoun[1,3,*], Ivana Ujević[2], Milad Fakhri[1], Romana Roje-Busatto[2,*], Céline Mahfouz[1], Sharif Jemaa[1] and Nikša Nazlić[2]

[1] *National Council for Scientific Research, National Center for Marine Sciences, Batroun, Lebanon*

[2] *Laboratory of Plankton and Shellfish Toxicity, Institute of Oceanography and Fisheries, Šetalište Ivana Meštrovića 63, Split, Croatia*

[3] *GEOMAR Helmholtz Centre for Ocean Research Kiel, Marine Biogeochemistry, Kiel, Germany*

Abstract: Marine biotoxins are naturally occurring chemicals produced by toxic algae. They can be found in seawater and can accumulate in various marine organisms, such as commercial seafood. When contaminated seafood is consumed, these biotoxins can cause poisoning in humans, with varying health consequences depending on the type and amount of toxins. The proliferation of biotoxin-producing algae in the marine environment has dire socio-economic and environmental consequences due to the contamination of water and seafood. Due to the number of factors related to human pressures and climate change impacts, the frequency of marine biotoxins' occurrence is increasing significantly globally, and in regional seas such as the Mediterranean Sea. In this chapter, we highlight Lebanon in the Eastern Mediterranean Sea, where marine biotoxins were recently studied. The results show for the first time the presence of lipophilic toxins and cyclic imines in marine biota, with values for okadaic acid, dinophysistoxin 1 and 2, pectenotoxin 1 and 2, yessotoxins and azaspiracids below the detection limit (LOD). Levels above LOD were detected for domoic acid (DA), gymnodimine (GYMb), and spirolides (SPXs) in some species/areas. Maximum levels of DA, GYM, and SPXs (3.88 mg DA kg-1, 102.9 µg GYM kg-1, 15.07 µg SPX kg-1) were found in the spiny oyster (*Spondylus spinosus*) in agreement with the occurrence of Pseudo-nitzchia spp, Gymndinium spp, and *Alexandrium spp*. DA was below the EU limit but above the lowest observed adverse effect level (0.9 µg g-1) for neurotoxicity in humans and below the acute reference dose (30 µg kg-1 body weight), both established by EFSA. Considering the lowest lethal dose (LD50) after administration of GYM and SPXs to mice, it is unlikely that there is a health risk due to exposure to these toxins from seafood consumption in Lebanon. Nevertheless, the chronic toxicity of DA, GYMs, and SPXs remains unclear, and the effects of repeated

* **Corresponding authors Abed El Rahman Hassoun and Romana Roje-Busatto:** National Council for Scientific Research, National Center for Marine Sciences, Batroun, Lebanon and Laboratory of Plankton and Shellfish Toxicity, Institute of Oceanography and Fisheries, Šetalište Ivana Meštrovića 63, Split, Croatia; E-mails: abedhassoun@cnrs.edu.lb and rroje@izor.hr

consumption of contaminated seafood need to be investigated. Because biotoxins have been detected in bivalves and commercial species, as well as other organisms in the marine trophic chain, it is evident that species other than bivalves should be monitored, and the spiny oyster (*S. spinosus*) may play the role of a sentinel species in biotoxin studies. A regular monitoring program is needed to provide reliable, accurate estimates of bloom toxicity and to investigate their potential impact on marine species and human health in Lebanon.

Keywords: Cyclic imines, Emergent pollutants, Lipophilic toxins, Lebanon, Mediterranean sea, Marine biotoxins, Marine biota, Public health, Seafood.

INTRODUCTION

Phytoplankton blooms have been known since the earliest human records (Boni, 1992; Zheng and Klemas, 2018). The global increase in these events has been notable since the 1980s (Smayda, 1990; Boni 1992, Vlamis and Katikou, 2015; Vilariño *et al.*, 2018) and has been attributed to favorable external conditions such as nitrogen/phosphorus resources, pH, and temperature (Stauffer *et al.*, 2020; Zhang *et al.*, 2020). More recently, these blooms have been attributed in part to the effects of ocean warming, marine heat waves, oxygen depletion, eutrophication, and pollution (Gobler *et al.*, 2017; Gobler *et al.*, 2021).

Of more than 70,000 phytoplankton species worldwide (Guiry, 2012), about 300 species can cause "red tides" (Hallegraeff *et al.*, 1995; Lindahl, 1998), and of these, more than 100 are producers of natural toxins that generate toxic episodes, known as Harmful Algal Blooms (HABs), that can be dangerous to humans and other organisms (Berdalet *et al.*, 2016).

According to Karlson *et al.* (2021), HABs can be divided into six main categories based on their adverse effects on the environment and/or human health: 1. Those that produce phytotoxins that accumulate in suspension feeders (bivalves); 2. Those that cause damage to respiratory mechanisms (fish gills) and/or feeding responses through toxin transfer, leading to mortality of fish and other marine life; 3. Blooms with high biomass that cause nuisance effects and/or lead to oxygen depletion; 4. Blooms disrupt the ecosystem and have multiple cascading effects on species interactions; 5. Those that produce aerosolized toxins that affect human respiratory health; and 6. Localized blooms of harmful benthic or epiphytic microalgae differ from planktonic HABs in habitat, mechanisms, and magnitude of adverse effects.

Phytoplankton cells form the base of the marine food chain and are an important food for filter-feeding bivalves and larval fish and crustaceans (Powell *et al.*, 1995; Cloern and Dufford, 2005). Consequently, HABs' toxins can be bioaccumulated and biomagnified in the marine trophic chain (Orellana *et al.*, 2017), and may be detrimental to plants, animals, people, and ecosystems (Harrness, 2005; Costa *et al.*, 2017). Thus, HABs can generate several socio-economic implications (Visciano *et al.*, 2016; Nwankwegu *et al.*, 2019; Zhongming *et al.*, 2021; Corriere *et al.*, 2021) which often depend on the size, severity, timing, and duration of the event (Qiao and Saha, 2021).

Biotoxins can be divided into hydrophilic and lipophilic molecules that can cause different symptoms: water-soluble toxins that cause Paralytic (PSP) and Amnesic Shellfish Poisoning (ASP), while liposoluble toxins cause Diarrhetic (DSP) and Neurotoxic Shellfish Poisoning (NSP) (FAO/IOC/WHO, 2004; Visciano *et al.*, 2016). Skin contact with contaminated water, inhalation of aerosolized biotoxins, or direct consumption of contaminated seafood can result in the effects of HABs on human health (Visciano *et al.*, 2016; Sonaka *et al.*, 2018).

Lipophilic Toxins (LTs)

Based on their polarity, marine biotoxins can be classified as hydrophilic, lipophilic (LTs), or amphiphilic (Alarcon *et al.*, 2018).

LTs are toxic metabolites from phytoplankton (*dinoflagellates*) isolated from different bivalve species (Draisci *et al.*, 1996) and classified into different classes (Liu *et al.*, 2019): Okadaic acid (OA), Dinophysistoxins (DTXs) and Azaspiracids (AZAs) which cause Diarrhetic shellfish poisoning (DSP) (Vale and Sampayo, 2002) and are considered tumor promoters (Fujiki and Suganuma, 1993). These also can cause pathological changes in the liver, pancreas, thymus, and spleen of mice (Ito *et al.*, 2000); while Pectenotoxins (PTXs) and Yessotoxins (YTXs) have not been shown to cause diarrhetic symptoms following intoxication (EFSA, 2008, 2009; Vlamis and Katikou, 2015; Ferron *et al*, 2016), but Domoic Acid (DA) is a potent neurotoxin responsible for Amnesic shellfish poisoning (ASP) that causes damage to the central nervous system (Gago-Martínez and Rodríguez-Vázquez, 2000; Diogène, 2017).

The Cyclic Imines (CIs)

With the discovery of new detection methods, the toxin groups are constantly updated and new toxins are identified and classified as "emerging toxins". Cyclic Imines (CIs), Palytoxin (PlTX), and Ciguatoxin (CTX) are examples whose appearance in the environment may be due to climate change affecting the distribution of phytoplankton species (EFSA, 2009, 2010a, 2010b). CIs, discovered in Canada in 1991 (Munday, 2008), are associated with algal blooms and shellfish contamination and are neurotoxins, antagonists of nicotinic receptors that affect the central nervous system (Otero *et al.*, 2011). CIs are macrocyclic compounds with 14 to 27 carbon atoms and two highly conserved moieties: the cyclic imine group (mainly a spiroimine) and the spiroketal ring system (Vlamis and Katikou, 2015; Molgó *et al.*, 2017). CIs are exemplified by 40 molecules that differ in the number of their rings: 5-membered (portimines), 6-membered (gymnodimines, spiroprorocentrimines, prorocentrolides), or 7-membered rings (spirolides, pinnatoxins, pteriatoxins), all of which are considered essential components for bioactivity (Stivala *et al.* 2015 ; Molgó *et al.*, 2017). They were grouped together based on their common imine group (part of a cyclic ring), which is responsible for their pharmacological and toxicological activity, and on their rapid acute toxicity in the mouse intraperitoneal bioassay (EFSA, 2010; Otero *et al.*, 2011; Reverté *et al.* 2014).

CIs are produced by marine dinoflagellate microorganisms such as *Karenia selliformis* and *Alexandrium ostenfeldii*/*A.* peru*via*num, which are associated with the biosynthesis of gymnodimines (GYMs) and spirolides (SPXs) (Seki *et al.*, 1995; Cembella *et al.*, 2000; Touzet *et al*, 2008; Salgado *et al*, 2015), while *Vulcanodinium rugosum* produces pinna toxins and portimines (Nezan and Chomerat, 2011; McCarthy *et al*, 2015; Molgó *et al*, 2017), but prorocentrolides were isolated from *Prorocentrum lima* (Torigoe *et al*, 1988), spiro- and prorocentrimines are probably produced by Prorocentrum species (Lu *et al*, 2001).

Poisoning by Marine Biotoxins

Worldwide Cases of Poisoning

According to García *et al.* (2018), an average of 60,000 people worldwide are poisoned by marine biotoxins each year. Anderson (1989), Quilliam (1993), Anderson (1994, 1997), and Okaichi (2004) have also noted an increase in the frequency and geographic distribution of poisonings worldwide. This high number of cases is attributed to the fact that these toxins are resistant to high temperatures, cannot be smelled, and contaminated seafood appears visually normal (Sobel and Painter, 2005). For example, in Prince Edward Island (Canada) in 1987, 153 deaths were attributed to DA poisoning from the consumption of clams, mussels, oysters, scallops, squid, sardines, anchovies, crabs, and lobsters (Quilliam and Wright, 1993; Lopez-Rivera *et al.*, 2009); and mild memory loss associated with DA has also been observed in Native Americans (Grattan *et al.*, 2016a, b). Seafood poisoning affects not only humans, but also various marine and coastal species that ingest the toxin, such as marine mammals, seabirds, fish, and others (Reyes-Prieto, 2009). Consequently, marine biotoxins do not only threaten ecosystem health, but also affect local economies due to their negative impacts on tourism, recreation, and aquaculture (Morgan ., 2009; *et al* García *et al.*, 2016), resulting in economic losses of approximately $82 million (NOAA, 2017; Anderson *et al.*, 2021).

Poisonings in the Mediterranean Sea

Over the past 50 years, a number of HABs have occurred in different areas of the Mediterranean Sea (McNamee *et al.*, 2016). The most recent HAB events were observed in Liguria, Italy, where more than 200 people (tourists and swimmers) experienced various symptoms of illness due to blooms of Ostreopsis ovata (Totti *et al.*, 2010; Ferrante *et al.*, 2013). Also in Italy, mussels contaminated with OA were found in a cultivation area in the Gulf of Trieste, causing poisoning in more than 300 people in 2010 because they were contaminated with various Dinophysis spp. in amounts up to 20 times the legal limit (Bacchiocchi *et al.*, 2015). In addition, several cases of PSP have been reported in different Mediterranean regions such as France, Italy, Morocco, Spain, and Tunisia (Fonda, 1996; Tahri Joutei, 1998; Rhomdane *et al.*, 1998; Taleb *et al.*, 2001; Lilly *et al*l., 2002; EU-NRL, 2002), and mild human ASP poisonings have occurred in Spain, France, Greece, and Italy (Friedman *et al.*, 2008). Although DSP toxins have been found in harvested shellfish in Croatia, no health problems due to the consumption of poisoned seafood have been recorded there (Orhanovic *et al.* 1996), but severe outbreaks have affected several thousand people in other Mediterranean countries such as France, Greece, and Spain (Belin, 1993; Van Egmond *et al.*, 1993; Durborow, 1999; EU-NRL, 2001; EU-NRL, 2002; FAO, 2004; Ferrante *et al.*, 2013; Costa *et al.*, 2017).

DA occurrence is associated with *Pseudo-nitzschia spp.* (Sobel and Painter, 2005 ; Ujević *et al.*, 2010 ; Moschandreou *et al.*, 2012) and *Nitzschia bizertensis* blooms (Bates *et al.*, 2018). This biotoxin was found in wild and cultured bivalve species in Croatia (Ujević *et al.*, 2010), France (Amzil *et al.*, 2001), Greece (Kaniou-Grigoriadou *et al.*, 2005), Italy (Ciminello *et al.*, 2005), and Portugal (Vale and Sampayo, 2001), as well as in Mediterranean lagoons in Tunisia (Sahraoui *et al*, 2012; Bouchouicha-Smida *et al.*, 2015). Concentrations of DA were below the regulatory limit with a maximum of 6.5486 µg g-1 (detection limit of 0.1025 µg g-1) in mussel samples from the Croatian Adriatic region (Ujević *et al*, 2010) and below 50 µg g-1 of body weight in France in 2002 (EU-NRL, 2002).

GYMs have been detected in bivalve species, mainly mussels, and clams, in various Mediterranean regions. In Croatia, GYMs were determined at concentrations below the limit of quantification [< 15 µg kg-1] (Gladan *et al.*, 2011; Ujević *et al.*, 2019), while in Greece, concentrations ranged from trace levels to 66 µg kg-1 within the framework of the Greek HAB monitoring program with the highest concentrations resulting in positive MBA tests (Vlamis and

Katikou, 2015), but in Tunisia, GYMs were detected for the first time in the Mediterranean and attributed to blooms of the toxic dinoflagellate *Karenia selliformis* (Biré *et al*., 2002; EFSA, 2010c; Ben Naila *et al*., 2012).

SPXs were also detected in Italy (Pigozzi *et al*., 2008), Spain (Villar González *et al*., 2006), and Greece (Katikou *et al*., 2012), at concentrations ranging from trace to 118 μg kg-1, with the highest levels resulting in positive MBA tests (Vlamis and Katikou, 2015). Levels of SPX (13-desmethyl spirolide C) in bivalve species ranged from 0.65 to 5.90 μg kg-1 in Croatian bivalve species (Ujević *et al*., 2019) and from 2 to 60 μg kg-1 in mussels and oysters from Catalonia, Spain (Amzil *et al*., 2007; García-Altares *et al*., 2014).

HABs have been detected on many occasions in Lebanese coastal waters, reflecting an anthropogenically enforced imbalance of the marine environment due to human activities (Abboud-Abi Saab *et al*., 2006, 2008a, 2008b; Abboud-Abi Saab and Hassoun, 2017). Except for a single recent study (Hassoun *et al*., 2021), there is no previous study quantifying biotoxins in marine species, either in Lebanon or in the Levantine sub-basin in general. Therefore, this chapter aims to describe in more detail the results recently published in the baseline study (Hassoun *et al*., 2021) to shed light on the biotoxin profile of different marine species collected in the Lebanese coastal region in the Eastern Mediterranean Sea as a case study.

STUDY AREAS AND METHODOLOGY

Areas and Species Studied

Marine species were sampled during the winter season, between December 2019 and February 2020, in the coastal areas of three Lebanese cities: Beirut (the capital and largest city), Tripoli (the second largest city in the north of the country), and Tyre (located in the south of the country and less exposed to anthropogenic influences compared to the other two cities); (Fig. **1**).

These wild species (82 individuals) were collected either directly from the sea or the fish market shortly after capture. Details of the species collected, sampling dates, and locations are given in Table **1**.

The 82 individuals sampled belong to four different species, including one gastropod, two mollusks, and one fish species. The species can be briefly described as follows:

Phorcus turbinatus is a gastropod known as a sea snail with a spirally coiled solid and thick shell that inhabits the rocky coasts of the Mediterranean Sea (Templado and Rolán, 2012). Phorcus are important algal feeders that play an important role in regulating the ecological balance in their habitat and are often used as biological indicators for assessing the impact of anthropogenic impacts on the coastal ecosystem (Donald *et al*, 2012; Henriques *et al*., 2017; Sousa *et al*., 2018) due to their sedentary nature, wide distribution, year-round abundance, and ability to accumulate chemicals such as heavy metals (Rainbow, 1997; Wang and Ke, 2002; Grupta and Singh, 2011).

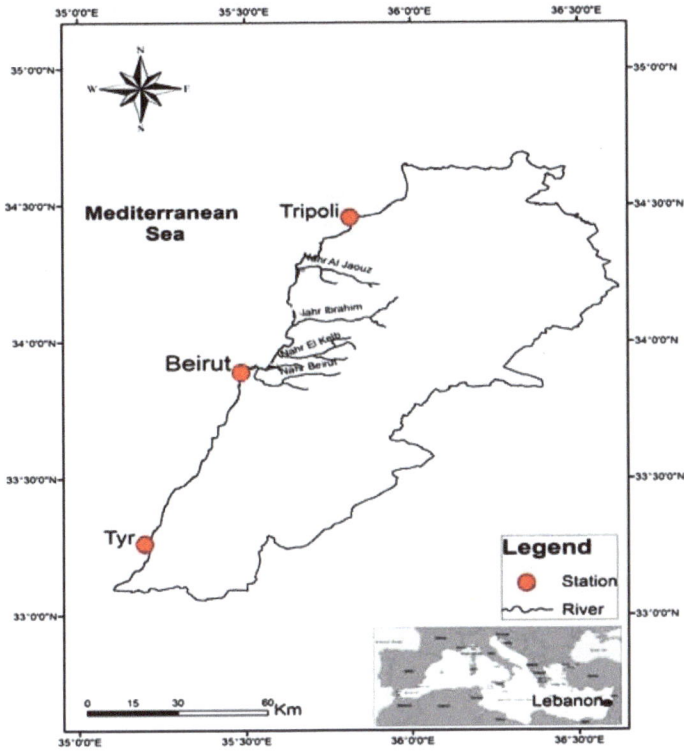

Fig. (1). Map illustrating the study areas.

Table 1. Sampling dates, locations, and number and characteristics of species examined in this study.

Date	Station	Longitude	Latitude	Species	Number of Individuals	Remarks	Trophic Level	Edibility
03-12-2019	Tripoli	35°48'27"	34°26'57"	P. rustica complex	12	Collected from the sea	Grazer	Occasional
03-12-2019	Beirut	35°28'32"	33°54'5"	P. rustica complex	12	Collected from the sea	Grazer	Occasional

(Table 1) cont.....

Date	Station	Longitude	Latitude	Species	Number of Individuals	Remarks	Trophic Level	Edibility
14-01-2020	Tyre	35°12'10"	33°16'2"	P. rustica complex	12	Collected from the sea	Grazer	Occasional
14-01-2020	Beirut	35°28'34"	33°54'5"	P. turbinatus	12	Collected from the sea	Grazer	Occasional
14-01-2020	Tyre	35°12'10"	33°16'2"	P. turbinatus	11	Collected from the sea	Grazer	Occasional
14-01-2020	Beirut	35°28'34"	33°54'5"	S. rivulatus	2	Seafood market	2nd lvel predator	Frequent
14-01-2020	Tyre	35°12'10"	33°16'2"	S. rivulatus	1	Seafood market	2nd lvel predator	Frequent
19-02-2020	Tripoli	-	-	*S. spinosus*	11	Seafood market	Filter-feeder	Frequent
19-02-2020	Tyre	-	-	*S. spinosus*	9	Collected from the sea	Filter-feeder	Frequent

Table 2. Quality parameters for each lipophilic toxin, including Precursor Ion (Prec ion), Product ion (Prod ion), Repeatability (RSD; Relative Standard De*via*tion), LOD (Limit of Detection), LOQ (Limit of Quantification), and Recoveries.

Toxin	Prec ion (m/z)	Prod ion (m/z)	Repeatability (RSD, %)	LOD ($\mu g\ kg^{-1}$)	LOQ ($\mu g\ kg^{-1}$)	Recovery (%)
OA	803.5	255.1 113.0	6.37	2.20	7.05	98
DTX1	817.5	255.1 113.0	9.79	4.19	13.40	107
DTX2	803.5	255.1 113.0	4.41	3.64	11.66	105
YTX	1141.5	1061.7 855.5	10.86	4.79	15.33	105
homoYTX	1155.5	1075.5 869.5	9.26	8.38	26.83	103
PTX2	876.5	823.4 213,2	10.33	3.05	9.75	94
AZA-1	842.5	824.5 806.5	8.11	1.77	5.68	89
AZA-2	856.5	838.5 820.5	6.34	2.34	7.49	102

(Table 2) cont.....

Toxin	Prec ion (m/z)	Prod ion (m/z)	Repeatability (RSD, %)	LOD (µg kg⁻¹)	LOQ (µg kg⁻¹)	Recovery (%)
AZA-3	828.5	810.5 792.5	7.08	2.66	8.50	93
SPX	692.2	444.0 163.8	5.27	2.05	6.56	104
GYM	508.0	490.1 392.4	4.83	2.36	7.54	104
DA	312.2	266.0 248.0	7.88	102.5	341.6	67

Patella rustica complex is a limpet found on rocky coasts of the intertidal and subtidal zones. This bivalve mollusk is used as a good environmental bioindicator because it is sedentary, sessile, ubiquitous, easily collected, and able to accumulate and tolerate strong concentrations of exogenous substances of anthropogenic origin (Nakhlé *et al.*, 2006; Amer *et al.*, 2018). Patella rustica is also considered a sentinel organism in monitoring programs for coastal environments due to its ability to accumulate metal elements in its tissues (Bergasa *et al.*, 2007).

Spondylus spinosus is a spiny oyster, a bivalve mollusk of hard substrate (Bitar, 2014). This oyster is a Lessepsian immigrant that was first documented in the Eastern Mediterranean Sea in 1988 (Mienis *et al.*, 1993) and has since expanded its range northwest to Lebanon and Turkey (Çeviker and Albayrak, 2006; Katsanevakis *et al.*, 2009; Crocetta *et al.*, 2013). *Spondylus spinosus* is established on rocky seabeds at depths of 2 to 40 m and forms dense and strong populations. They can occur in both the infra-littoral and circa-littoral environments (Crocetta *et al.*, 2013). *S. spinosus* is commonly exploited locally as an edible seafood delicacy. The external morphology of *S. spinosus* can provide many useful features for evolutionary studies and can be used for biomonitoring studies in both natural and experimental contexts (Aksit *et al.*, 2013).

Siganus rivulatus, or marbled spinefoot is a Lessepsian species that emigrated from the Red Sea and entered the Mediterranean after the opening of the Suez Canal (Bariche, 2006; IUCN, 2009). It is a demersal fish that inhabits shallow waters and generally occurs in schools of fifty to hundreds of individuals (Ben Rais Lasram and Mouillot, 2009; Zenetos *et al.*, 2010). This species is herbivorous and feeds on available macrophytes depending on the season, preferring some species such as Ulva spp., Jania spp., and Halopteris spp. In Lebanon, this species is widely consumed by local residents and fishermen and is prevalent in artisanal fisheries (Barbour *et al.*, 2009).

METHODS OF ANALYSIS

Physico-chemical Parameters

Surface temperature, phosphates, nitrates, nitrites, and chlorophyll-a concentrations were measured monthly during the winter season at the study sites, as part of the National Monitoring Program of the Lebanese coast conducted by the CNRS-L. Temperature (T) was measured with an ordinary thermometer. Since salinity (S) was not measured during the study period (2019 winter season), the S value was averaged for the winter season (December-February) of the 2016-2019 period (n= 3, 4, 5 for Tyre, Tripoli, and Beirut, respectively).

Orthophosphates (P-PO4) were analyzed according to the method described by Murphy and Riley (1962), nitrites (N-NO$_2$) according to Bendschneider and Robinson (1952), and nitrates (N-NO$_3$) according to Strickland and Parsons (1968) with a small modification due to the use of ammonium chloride as an activator (Grasshoff, 1961). Total chlorophyll-a (Chl-a) was measured by filtering the samples at low pressure through a filter paper (Whatman GF/C) and extracting the pigments in 90% acetone for 24 hours under cold and dark conditions. The concentration was determined with a spectrophotometer according to the monochromatic method of Lorenzen (1967). Biomass is expressed as Chl-a amount per seawater volume (µg L-1; Abboud-Abi Saab and Hassoun, 2017).

Phytoplankton Analysis

Phytoplankton samples were collected several days after sampling the marine biota and immediately placed in Lugol's iodine solution at a final concentration of 0.5% to preserve them for species identification. Utermöhl sedimentation method (Utermöhl, 1958) was used, in which 100 ml of homogeneous samples were settled in a 25 mm diameter sedimentation chamber for 48 hours. The bottom of the chamber was examined with a Wild M 40 phase-contrast inverted microscope, and cells were counted at 40X magnification to ensure better identification of *dinoflagellates*. Dominant cells were identified at the species level, while taxonomic groups such as small pennate diatoms, flagellates, and naked *dinoflagellates* were counted in groups.

Biotoxin Analysis

The lipophilic toxins analyzed in this study were okadaic acid (OA), dinophysistoxin 1 and 2 (DTX1, 2), pectenotoxin 1 and 2 (PTX1, 2), yessotoxins (YTXs), azaspiracids (AZAs), gymnodimine (GYMb), spirolides (SPXs), and domoic acid (DA). All of these toxins were analyzed in homogenized tissues. For the analysis of LTs and CIs, 2.00 g ± 0.05 g of homogenized whole-body shellfish tissue (for gastropods and bivalves) and muscle tissue (for fish) were extracted before injection into the LC-MS / MS (EURLMB- Harmonized SOP, 2015). For extraction, 9.0 ml of 100% methanol was added, then the sample was mixed with the shaker for 3 minutes and then centrifuged at 2000 g or higher for 10 minutes at approximately 20°C. After two centrifugations, the supernatants of the two resulting extracts were combined to obtain an extract of 20 ml containing 100% methanol. The extract was then filtered through a dry methanol compatible 0.45-µm syringe filter.

To detect and quantify the total content of some toxins (such as OA and DTX toxins), alkaline hydrolysis was required prior to LC-MS/MS analysis to convert the acylated toxins present in the sample to the parent toxins (*i.e.*, OA, DTX1, and DTX2) (Mountfort *et al.*, 2001). The positive ion mode of the LC-MS/MS method was used for the detection of AZA1, AZA2, AZA3, PTX1, PTX2, SPX, GYM, and DA (domoic acid), while the negative mode was used for the detection of YTX, homoYTX, 45 OH-YTX, 45 OH-homoYTX, OA, DTX1, DTX2, and acylated esters of OA and DTXs. Further information on reagents, steps, and procedures can be found in the EURLMB-Harmonised SOP (2015). The LC-MS/MS analysis was performed using a triple-quadrupole mass spectrometer (Agilent Technologies 6410) equipped with an electrospray ionization source. Chromatographic separation was performed using a 5µm Poroshell C18, 50×2.1mm Agilent column maintained at 30°C. The hydrolysis procedure proposed by EURLMB (2015) was applied to investigate the total content of OA and DTX toxins. The detection limits (LOD) were: 1.77 for AZA1, 2.34 for AZA2, 8.50 for AZA3, 2.20 for OA, 4.19 for DTX1, 3.64 for DTX2, 3.05 for PTX2, 4.79 for YTX, 8.38 for homoYTX, 2.36 for GYM, and 2.05 µg kg-1 for SPX.

RESULTS

Biotoxins Present

The results show that all levels of okadaic acid (OA), dinophysistoxin 1 and 2 (DTX1, 2), pectenotoxin 1 and 2 (PTX1, 2), yessotoxins (YTXs), azaspiracids (AZAs) were below their detection limits (LOD). Only domoic acid (DA), gymnodimine (GYMb), and spirolides (SPXs) were found above the detection limit in some species/areas.

Amnesic Shellfish Poisoning Toxin

Domoic Acid (DA)

Amnesic Shellfish Poisoning toxin, DA, was determined in our samples (Fig. **2**). All samples from Phorcus, Patella, and Siganus had concentrations below the detection limit (< 0.1025 mg kg-1). However, only 18 and 33% of *S. spinosus* samples from Tripoli and Tyre, respectively, had concentrations below the detection limit, while the other 82 and 67% (respectively) had concentrations above the DL with an average of 2.6 ± 1.08 mg kg-1 for *S. spinosus* samples from Tripoli, six times higher than those from Tyre (0.38 ± 0.17 mg kg-1). DA Concentrations in these samples ranged from a minimum of 0.1516 mg kg-1 for an *S. spinosus* sample from Tyre to a maximum of 3.88 mg kg-1 for an *S. spinosus* sample from Tripoli (Fig. **3**). All results from DA are below the maximum level for consumption (20 mg kg-1) established in the EU regulation [Regulation (EC) No. 853/2004-30/04/2004 Regulation of the European Parliament and of the Council of 29 April 2004 laying down specific rules on the hygiene of foodstuffs].

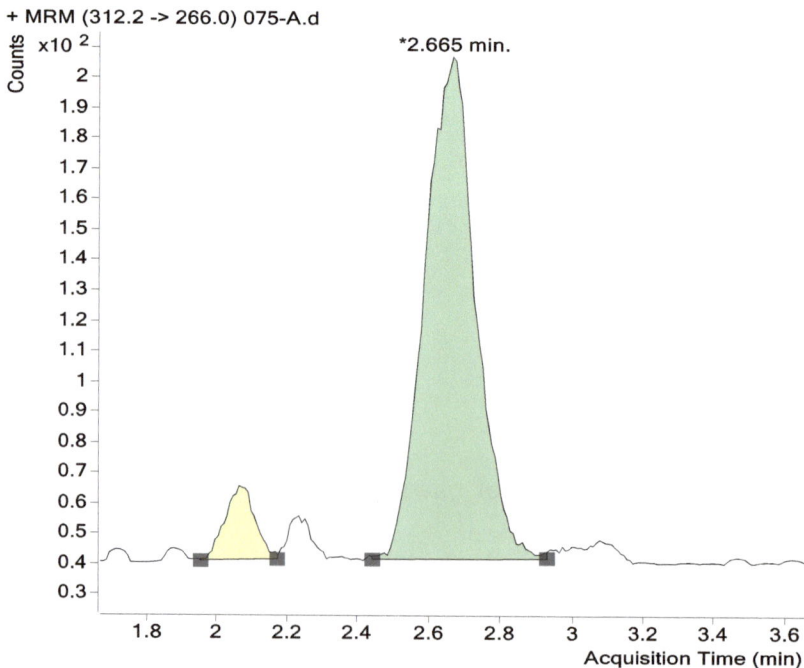

Fig. (2). Multiple reaction monitoring (MRM) chromatogram confirming the presence of DA and its retention/acquisition time.

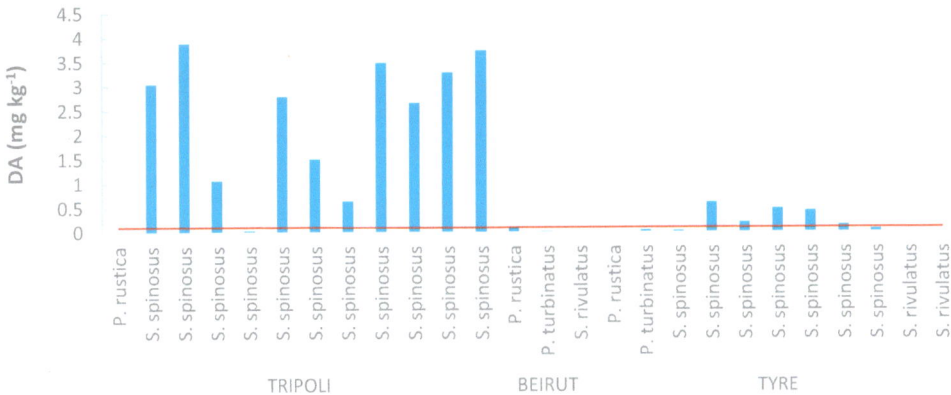

Fig. (3). DA concentrations in the different studied species collected in Tripoli, Beirut, and Tyre. The red line corresponds to the DA detection limit (0.1025 mg kg−1).

Cyclic Imines (CIs)

Gymnodimines (GYMs)

The main peak of GYM corresponds to GYMb, and GYMs hereafter refers to both GYMb and its derivative (the other peak). GYMs were detected in 66% of our samples (Fig. **4**) at concentrations above the detection limit (2.36 µg kg-1), whereas the concentrations of GYM varied from 3.33 µg kg-1 in S. rivulatus (Tyre) to 102.9 µg kg-1 in *S. spinosus* (Tripoli) and were below the detection limit in all gastropod samples (Fig. **5**). The spiny oyster *S. spinosus* is the species that had the highest GYM concentrations, averaging 56 ± 27 µg kg-1 from the Tripoli samples, 1.5 times higher than the average of the samples from Tyre (36 ± 11 µg kg-1). The second highest GYM load was exhibited by the limpet *P. rustica* complex with the highest average values from Tyre (26.9 µg kg-1), Beirut (26.8 µg kg-1), and Tripoli (8.7 µg kg-1). The lowest GYM concentration was found in the fish *S. rivulatus* collected in Tyre (3.33 µg kg-1).

Spirolides (SPXs)

SPX (13-desmethyl-spirolide C) was detected in our samples (Fig. **6**), with concentrations above the detection limit (2.0484 µg kg−1) measured only in the spiny oyster *S. spinosus*. SPX concentrations ranged from a minimum of 2.18 to a maximum of 15.07 µg kg−1 in oysters from Tripoli and Tyre respectively (Fig. **7**) and were found mainly in oysters from Tyre (67%) while only 27% of *S. spinosus* samples from Tripoli were contaminated with this toxin. The average SPX concentration in *S. spinosus* taken from Tyre (7.54 ± 5 µg kg−1) is 1.86 times higher than the average of samples from Tripoli (4.05 ± 2 µg kg−1).

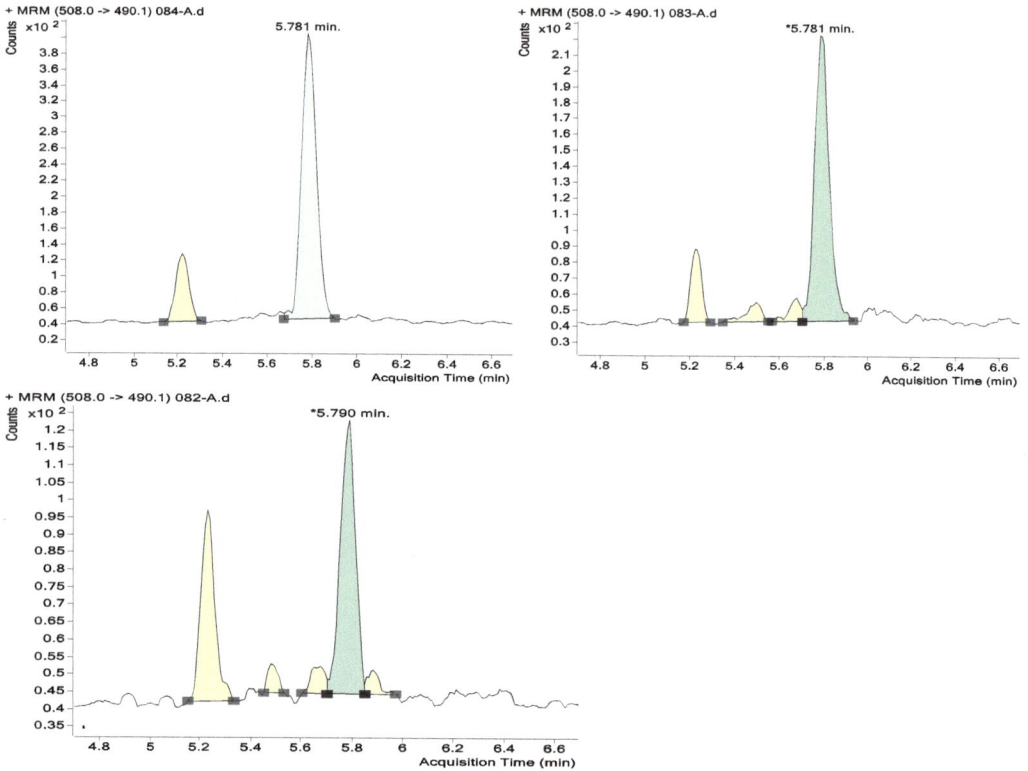

Fig. (4). Multiple reaction monitoring (MRM) chromatogram confirming the presence of GYMs and its retention/acquisition time.

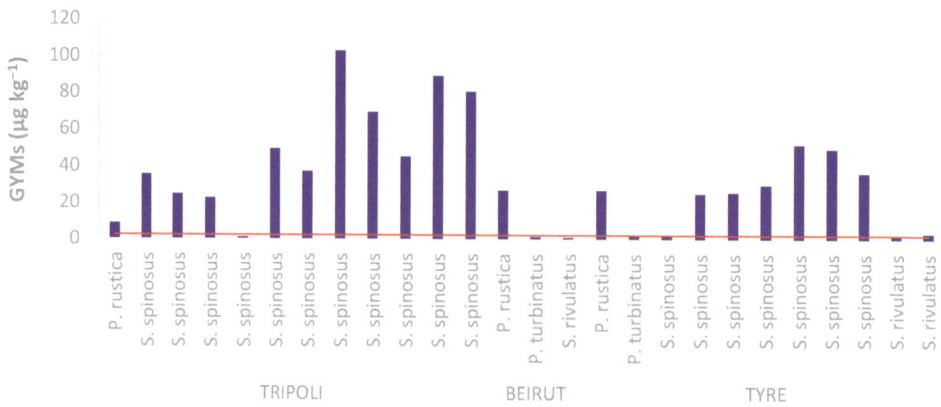

Fig. (5). GYMs concentrations in the different studied species collected in Tripoli, Beirut, and Tyre. The red line corresponds to the GYMs detection limit (2.3568 µg kg−1).

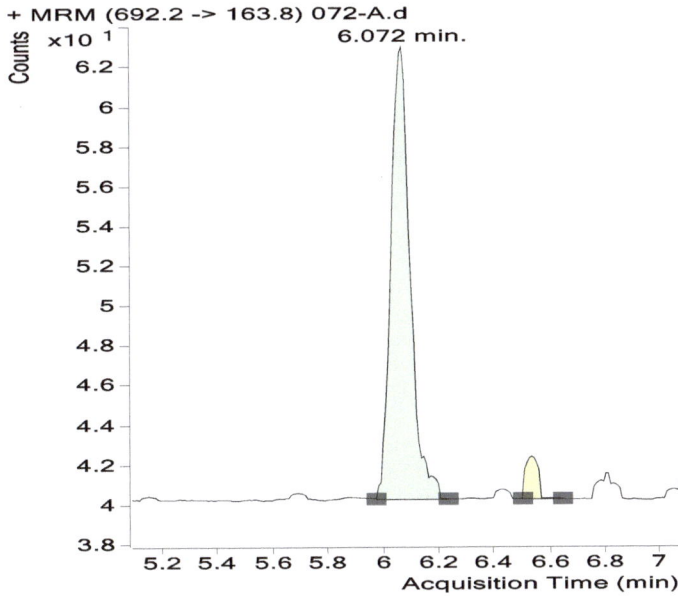

Fig. (6). Multiple reaction monitoring (MRM) chromatogram confirming the presence of SPX and its retention/acquisition time.

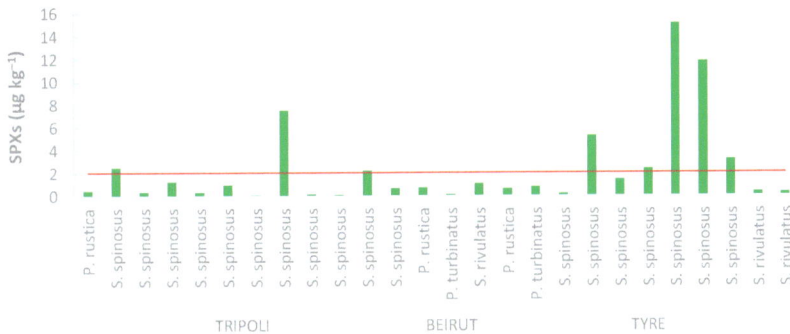

Fig. (7). SPXs concentrations in the different studied species collected in Tripoli, Beirut, and Tyre. The red line corresponds to the SPXs detection limit (2.0484 µg kg−1).

Physico-chemical Parameters

During the study period (winter season), the highest hydrographic parameters (temperature and salinity) were measured in Beirut. The lowest temperature was recorded in Tyre, and comparable salinity levels were measured in both Tripoli and Tyre (Table **3**).

Table 3. Average values of physico-chemical parameters in surface waters (~ 0.5 m) during winter in the study areas.

Location	Temperature	Salinity*	Phosphates	Nitrates	Nitrites	Chl-a	Pheopigments
Units	°C		μmol L-1	μmol L-1	μmol L-1	mg m3	mg m3
Tripoli (n=2)	20.05 ± 0.95	38.49 ± 0.01	0.3 ± 0.07	2.2 ± 0.8	0.164 ± 0.035	0.075 ± 0.05	0.09 ± 0.05
Beirut (n=2)	21.25 ± 0.25	38.59 ± 0.54	0.07 ± 0.009	1.18 ± 0.15	0.125 ± 0.016	0.21 ± 0.07	0.22 ± 0.03
Tyre (n=1)	18.5	38.47 ± 0.22	0.22	5.09	0.28	0.55	1.92

*Salinity values were averaged for the winter season (December-February) of 2016-2019 (n= 3, 4, 5 for Tyre, Tripoli, and Beirut, respectively).

The nutrients' concentrations indicate that the ecological quality status of the three study areas can be classified as "good" based on the criteria recommended by Karydis (2009) for the coastal waters of the Eastern Mediterranean. Tripoli had the highest orthophosphate concentrations (4 and 1.4 times higher than the values in Beirut and Tyre, respectively), while the highest nitrate and nitrite concentrations were measured in Tyre. Nitrate concentrations in Tyre were 4 and 2 times higher than the values of Beirut and Tripoli, respectively, while nitrite concentrations in Tyre were 2 and 1.7 times higher than the values in Beirut and Tripoli, respectively (Table **3**).

Consistent with nutrients, the highest primary production was also obtained in Tyre with the highest chlorophyll-a and pheopigment concentrations, 7 and 2.5 times higher than Chl-a levels in Tripoli and Beirut, respectively (Table **3**).

PHYTOPLANKTON POPULATIONS

The results of the phytoplankton analysis show the presence of the species producing biotoxins found in this study, namely *Pseudo-nitzschia* spp., *Gymnodinium spp.*, and *Alexandrium* spp. as producers of domoic acid, gymnodimines, and both gymnodimines and spirolides, respectively (Fig. **8**). A complete list of the species can be found in Table **4**.

Table 4. Counted phytoplankton species at each site on 02.03.2020 and 03.04.2020.

Species	Tripoli	Beirut	Tyre
Diatoms	37485	91835	121635
Asterionella japonica	510	-	713
Bacteriastrum spp.	-	-	3567

(Table 4) cont.....

Species	Tripoli	Beirut	Tyre
Cerataulina spp.	-	-	2854
Chaetoceros spp.	-	-	22472
Cocconeis sp.	-	-	1070
Climacosphenia sp.	-	-	1070
Cylindrotheca closterium	2295	2675	3567
Dactyliosolen fragilissimus	-	-	2140
Grammatophora spp.	-	1337	1427
Guinardia spp.	-	-	4994
Helicotheca sp.	1275	-	-
Leptocylindrus danicus	5355	-	2854
Licmophora sp.	2805	1783	2854
Pleurosigma sp.	-	-	1070
Pseudo-nitzschia spp.	8415	5350	4637
Skeletonema costatum	1020	-	-
Striatella delicatula	-	446	713
Striatella unipunctata	-	446	1070
Dinoflagellates	629046	105656	23900
Unidentified Dino.	-	61075	7491
Alexandrium spp.	255	892	1427
Gymnodinium spp.	281784	24965	7491
Gyrodinium spp.	345987	15603	7134
Pronoctiluca sp.	-	446	-
Prorocentrum lima	510	446	-
Prorocentrum micans	-	-	357
Protoperidinium sp.	-	2229	-
Ceratium sp.	510	-	-
Ciliates	510	1783	6064
Tintinnids	-	892	-
Euglena	-	892	-
Coccolithophores	255	5350	1070
Microphytoplankton	667296	206408	152669
Nanophytoplankton	189045	27640	28893

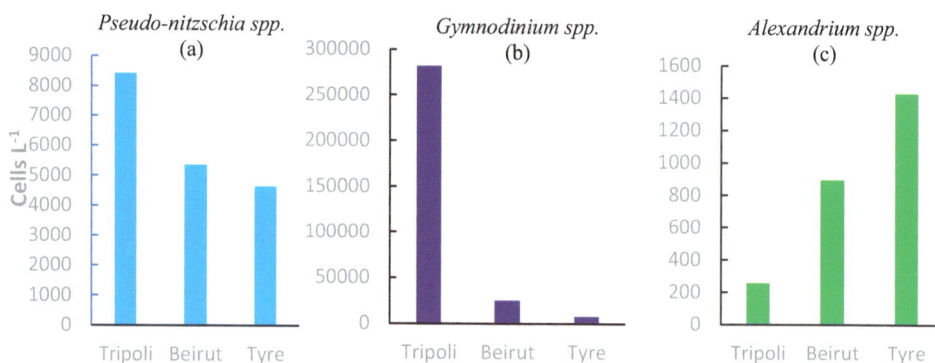

Fig. (8). Number of cells per litre (Cell L-1) for phytoplankton species that produce **a**) DA: Pseudonitzchia spp., **b**) GYMs: *Gymnodinium spp.*, and **c**) GYMs and SPXs: *Alexandrium spp.* in Tripoli, Beirut, and Tyre.

Pseudo-nitzschia spp. were found in concentrations of less than 4000 cells L-1 in the three areas studied (Fig. **8a**). The highest concentration of *Pseudo-nitzschia spp.* was found in Tripoli, consistent with the highest DA values measured in the same area (Fig. **3**). Although higher concentrations of *Pseudo-nitzschia spp.* were detected in Beirut than in Tyre, the DA values were higher in Tyre (Fig. **3**). *Gymnodinium spp.* was also found in all three stations, with concentrations above 104 cells L-1 in Tripoli (Fig. **8b**), consistent with the highest GYMs values in the same study area (Fig. **5**). However, the high GYMs values in Tyre are not consistent with the lowest *Gymnodinium spp.* concentrations found in Tyre (< 3500 cells L-1 ; Figs. **5** and **8b**). In addition, *Alexandrium spp.* were found in the three stations studied, with concentrations always below 1000 cells L-1 and the maximum detected in Tyre, which is consistent with the highest SPX concentrations also measured there. In Beirut, concentrations of *Gymnodinium spp.* were higher than in Tyre and of *Alexandrium spp.* higher than in Tripoli, which is not consistent with the lowest concentrations of GYMs and SPXs, both measured in Beirut (Figs. **7** and **8c**).

Microphytoplankton species were the dominant groups in the three areas studied (Fig. **9**), with *dinoflagellates* in both Tripoli and Beirut, where the highest temperature and salinity were measured, and diatoms as the dominant group in Tyre, where the lowest T and S values and the highest nitrate and nitrite values were measured (Fig. **9**; Table **3**). The predominant species were Gyrodinium spp. and *Gymnodinium spp.* both of which had their highest concentrations in Tripoli (Table **4**).

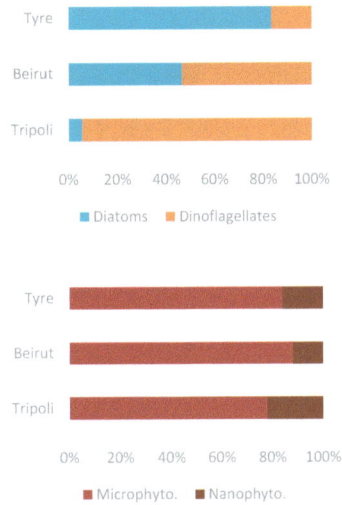

Fig. (9). Percentage of phytoplankton populations **a**) diatoms and *dinoflagellates*, **b**) total micro- and nano-phytoplankton in Tripoli, Beirut, and Tyre.

DISCUSSION

Nearly 70% of Lebanon's population are coastal residents, and major tourist and industrial activities are also located near or on the coast, resulting in the degradation of seawater quality. Lebanese coastal waters receive about 65% of the total wastewater through at least 53 major wastewater discharges along the Lebanese coast (CDR/LACECO, 2000; MOE/UNDP/ECODIT, 2011), which explains the relatively high nutrient concentrations in Tyre and Tripoli, as the samples collected could be influenced by human activities, *e.g.*, in nearby ports (Table **2**). In addition, rivers carry pollutants from agricultural runoff and sewage directly into the sea (MOE/UNDP/ECODIT, 2011). In addition, only 66% of homes and businesses were connected to an improved wastewater network in 2007, which explains Lebanon's 90th rank among 163 countries based on the Environmental Performance Index (EPI), indicating lower performance in terms of environmental sustainability (World Bank, 2011).

As a result, the narrow Lebanese coastline (~240 km) is highly urbanized and subject to significant anthropogenic pollution, primarily from sewage and industrial waste. As a result, algal blooms are also frequently observed in heavily polluted areas (Abboud-Abi Saab and Hassoun, 2017). Consequently, toxin producers (Fig. **8**) and/or HABs are present in many coastal areas, which explains the relatively high levels of biotoxins (Figs. **3 - 7**) accumulated in the marine biota studied.

Marine biotoxins in Lebanon and the Mediterranean Sea

The present study revealed that DA concentrations varied from 0.1516 to 3.88 μg g-1 in the spiny oyster *S. spinosus* (Fig. **3**). Although the concentration of DA has not previously been measured in this species, numerous studies have been conducted in other marine organisms that have found comparable or lower concentrations. In the Adriatic Sea, for example, the presence of DA was confirmed in wild populations of the rough cockle *Acanthocardia tuberculata* (0.17 to 0.77 μg g-1) and the smooth clam Callista chione (0.28 μg g-1) collected from the Croatian coast (Ujević *et al.*, 2019). However, most studies in the Mediterranean region targeted harvested species, mainly mussels and oysters, such as in the Bay of M'diq in Morocco, where a maximum DA of 4.9 μg g−1 was measured in the whole tissue of the sweet clam (C. chione) in spring 2007 (Rijat Leblad *et al.*, 2013), in Bizerte Lagoon of Tunisia with DA ranges of 0.13–0.86 μg g−1 in the mussel *Mytilus galloprovincialis* and 0.42–1.04 μg g−1 in the oyster *Ostrea edulis* (Bouchouicha-Smida *et al.*, 2015), in Languedoc-Southern France (*M. galloprovincialis*: 0.8 μg g−1 ; Amzil *et al.* 2001), in Spain and Portugal (M. galloprovincialis: c. 5 μg g−1 ; Vale and Sampayo 2001; Giménez *et al*, 2012), in Croatia (Flexopecten proteus, Pecten jacobaeus, *Ostrea edulis* and M. galloprovincialis: 0.1117-1.6567 μg g−1), whereas in Greece 83% to 95% of all sampled mussels (*M. galloprovincialis*) and venus clams (*Venus verucosa*) contained less than 1 μg g−1 in 2002 and 2003, respectively, with a maximum of 14.0 μg g−1 in mussels in 2002 and 4.2 μg g−1 in mussels and 5.6 μg g−1 in venus clams in 2003 (Kaniou-Grigoriadou *et al.*, 2005).

GYMs levels in our samples ranged from 3.33 to 102.9 μg kg-1 (Fig. **5**), a range that is higher than that found in many Mediterranean regions, such as in Greece, where concentrations in bivalves (mussels, venus clams, and hard clams) varied from trace to 66 μg kg-1 (Vlamis and Katikou, 2015), in Jamâa Ouled Ganem-Morocco, where GYMs in M. galloprovincialis were lower than 5.6 μg kg-1 (Haddouch *et al.*, 2017), in the eastern Adriatic Sea, where GYMs levels in mussel species (M. galloprovincialis and Mediterranean scallop, Pecten jacobaeus) from Šibenik Bay were below the detection limit (Gladan *et al.*, 2011), in the Dubrovnik area (Čustović *et al.*, 2012) and Mali Ston Bay (Arapov *et al.*, 2015), while GYMs values of 2.65-15.77 and 1.17-6.14 μg kg-1 were detected in A. tuberculata and C. chione, respectively, in the southern Adriatic Sea (Ujević *et al.*, 2019). Other studies did not find GYMs in their samples. For example, after analyzing several raw and processed commercial bivalves in eight European countries (including four Mediterranean countries: Italy, Portugal, Slovenia, and Spain), no GYMs were detected over a two-year period (Rambla-Alegre *et al.*, 2018). Similarly, no GYMs contamination was detected in clams and mussels collected in Ganzirri and Faro lakes (which are connected to each other and to the

Ionian and Tyrrhenian seas; Mattarozzi *et al*., 2019). In addition, plankton samples from the Ebro Delta (NW Med) contained very low GYM values (0.021 - 39.931 ng L-1; Busch *et al*., 2016). Otherwise, very high GYM concentrations were measured in the south of the Mediterranean, at least 5 times higher than the values obtained in our study (Fig. **5**). For example, in Tunisia, a range of 460-1290 µg kg-1 was found in the digestive glands of clams from the Tunisian coast (Marrouchi *et al*., 2009), and a maximum value of 2136 µg kg-1 was measured in clams from the Gulf of Gabes (Ben Naila *et al*., 2012).

SPX (13-desmethyl spirolide C) content in our samples ranged from 2.18 to 15.07 µg kg-1 in 67% and 27% of spiny oysters (*S. spinosus*) collected in Tyre and Tripoli, respectively (Fig. **7**). This range is higher than that observed in plankton samples from Alfacs and Fangar bays in the northwestern Ebro Delta (0.118 - 184.418 ng L-1; Busch *et al*, 2016), in *Galician M. galloprovincialis* (1.2-6.9 µg kg-1; Moreiras *et al*., 2019) and bivalve species (0.98-5.90 µg kg-1 in A. tuberculata and 0.65-2.14 µg kg-1 in C. chione) from the eastern Adriatic Sea (Ujević *et al*., 2019). In contrast, broader ranges of high SPX concentrations were found in mussels and oysters (2-60 µg kg-1) from Catalonia, Spain (Amzil *et al*., 2007; García-Altares *et al*., 2014), and in Greek shellfish with concentrations ranging from trace to 118 µg kg-1 (Katikou *et al*., 2012; Vlamis and Katikou, 2015). An interesting study showed that 9.4% of samples (n= 47) of raw and processed commercial shellfish from several European countries were contaminated with SPX, with concentrations ranging from 26 to 66 µg kg-1 (Rambla-Alegre *et al*., 2018). However, lower or no SPX levels were detected in other parts of the Mediterranean, such as Jamâa Ouled Ganem-Morocco (M. galloprovincialis; Haddouch *et al*., 2017). Also, all clams and mussels collected in Ganzirri and Faro lakes (associated with both the Ionian and Tyrrhenian seas) were free of SPX (Mattarozzi *et al*., 2019).

Sentinel Species for Monitoring Marine Biotoxins

Both oysters and mussels are considered bioindicators of ecosystem contamination and are used as sentinels of potential biotoxin integration (Turki *et al*., 2014). Analysis of the concentrations of DA in different marine species, as in the present study, is useful in determining organisms that are more susceptible to the accumulation of DA in their tissues than other species. Such organisms can therefore be considered early warning species indicating the presence of toxins in a marine ecosystem and can also be referred to as "sentinel" species. In fact, all of the above studies show that DA levels in oysters are higher than those measured in other marine animals such as bivalves. Clams can also hold DA for up to one year in the natural environment or several years after processing, canning, or freezing (Ferriss *et al*., 2017). In our study, the spiny oyster *S. spinosus* is the only species whose levels were above the detection limit DA (Fig. **3**), which is relatively high compared to the results of the above studies. These results are in agreement with other studies, for example, when mussels and oysters were collected simultaneously in the Bizerte lagoon (Tunisia), oysters contained higher DA levels than mussels, and this difference was attributed to the time required for excretion (Bouchouicha-Smida *et al.*, 2015). A statement that was also confirmed by Blanco *et al.* (2002), who found that mussels appeared to purge faster than oysters (DA). Although GYMs were found in several species in the present study, the spiny oyster *S. spinosus* had the highest levels (average 48.4 ± 24 µg kg-1; Fig. **5**). SPXs, on the other hand, were found only in *S. spinosus* (Fig. **7**). Among the marine organisms investigated in the present study, the spiny oyster *S. spinosus* may serve as a sentinel species that could be monitored regularly to detect the levels of DA, GYMs, and SPXs and to assess possible poisoning risks to humans in Lebanon.

The same trend was observed for heavy metals in seafood collected from the Lebanese coast. The bivalve *S. spinosus* had the highest concentrations of cadmium (Cd), lead (Pb), and arsenic (As) compared to fish species (Siganus rivulatus, Lithognathus mormyrus, and Etrumeus teres) and shrimp (Marsupenaeus japonicus) (Ghosn *et al.* 2019). In phytoplankton, a clear positive effect of low-dose toxic substances on organisms has been documented (Zhang *et al.*, 2020), which is considered a stimulatory effect of low-concentration environmental pollutants such as polycyclic aromatic hydrocarbons. This phenomenon, also known as "hormesis" should be evaluated for a sentinel species such as the spiny oyster, which can bioaccumulate different types of emerging pollutants. These results suggest that the spiny oyster can be used as a sentinel species to monitor different pollutants simultaneously present in the marine environment and to assess potential impacts on consumer health.

Biotoxin Producers

What Do We Know?

DA production is mainly attributed to the presence of *Pseudo-nitzschia spp.* whose bloom events preceded high DA concentrations in mussels, as demonstrated on the Croatian coasts of the eastern Adriatic (Ujević *et al.*, 2019). The genus Pseudo-nitzschia includes cosmopolitan species found worldwide (Hasle, 2002; Trainer *et al.*, 2008; Churro *et al.*, 2009; Loureiro *et al.*, 2009; Lundholm *et al.*, 2010) and in several Mediterranean areas (Quiroga, 2006; Congestri *et al.*, 2008; D'Alelio *et al*, 2008; Loureiro *et al.*, 2009; Sahraoui *et al.*, 2009; Amato *et al.*, 2010), including Lebanese waters, where *Pseudo-nitzschia delicatissima* is one of the dominant species in some areas and can account for 7 to 51% of the total microphytoplankton populations (Abboud-Abi Saab and Hassoun, 2017). During the winter season (December to February), the collection period of our samples, the proportion of P. delicatissima usually varies between 7 and 25%, depending on the station, and the highest concentration is reached during the decline in temperature and salinity (November to May), especially at stations influenced by terrestrial inputs such as rivers and sewage (Abboud-Abi Saab and Hassoun, 2017). In our phytoplankton samples, *Pseudo-nitzschia spp.* represents 1, 2.5, and 3% of the total microphytoplankton community in Tripoli, Beirut, and Tyre, respectively (Fig. **8a**), with the highest abundance in Tyre, where T and S show the lowest values (Table **2**). These results suggest that the high values of DA are not correlated with the occurrence of *Pseudo-nitzschia spp.* bloom during the study period. The same conclusion was highlighted in other studies where it was found that the maximum concentration of DA did not coincide with the bloom of this species and that other unidentified species could be responsible, as high levels above the legal limit were measured in soft tissues of Pecten maximus during the persistent low abundance of *Pseudo-nitzschia spp.* (James *et al.*, 2005). This has been documented in several areas of the Mediterranean Sea, *e.g.*, the Ebro Delta (Busch *et al.*, 2016), Arenys de Mar in 2008, and Vilanova in 2010 (both on the Catalan coast; Giménez Papiol *et al.*, 2013), as well as in the eastern Adriatic Sea, where potentially hazardous DA concentrations can accumulate in some bivalve species even when *Pseudo-nitzschia spp.* is present in low abundance (Ujević *et al.*, 2019). The same was found in the Bizerte lagoon (Tunisia), where the production of DA was primarily attributed to another species: *Nitzschia bizertensis* (Bouchouicha-Smida *et al.*, 2015). However, a good correlation between high DA levels and *Pseudo-nitzschia spp.* was also observed in other areas of the Mediterranean, *e.g.*, near the Krka estuary (central Adriatic Sea), where a correlation between high DA concentrations and high abundance of *Pseudo-nitzschia spp.* ($> 1.0 \times 106$ cells L-1) coincided with temperature and salinity minima in February (Ujević *et al.*, 2019). A similar correlation was also found in the Bay of Banyuls-sur-Mer in the northwestern Mediterranean during spring (Quiroga, 2006), along the coast of the Lazio, Middle Tyrrhenian Sea (Congestri *et al.*, 2008), and in the eastern Adriatic Sea (Ujević *et al.*, 2010).

The production of GYMs is mainly attributed to the dinoflagellate *Karenia selliformis*, previously identified as Gymnodinium selliforme (McKenzie *et al.*, 2002; Haywood *et al.*, 2004 Kremp *et al.*, 2014). *Gymnodinium spp.* is widely distributed in the Mediterranean Sea (Reñé *et al.*, 2011 and references therein), including the Levantine Sea (Siokou-Frangou *et al.*, 1999). In Lebanon, *Gymnodinium spp.* is among the dominant phytoplankton species in coastal waters, ranging between 6 and 29% between October and December, with the highest concentrations measured at stations with river discharges (Abboud-Abi

Saab and Hassoun, 2017). Also, our phytoplankton analysis showed that *Gymnodinium spp.* account for 5, 12, and 42% of the total microphytoplankton populations in Tyre, Beirut, and Tripoli, respectively (Figs. **8b**, **9**, and **10**). These results explain the high GYMs levels measured in Tripoli (Fig. **5**) and are consistent with many studies around the Mediterranean that have found a correlation between the presence of GYMs and a high abundance of *Gymnodinium spp.* such as Gladan *et al.* (2011), who demonstrated the presence of GYMs in the eastern Adriatic 12 days after the occurrence of a *Gymnodinium spp.* bloom (7.1 × 105 cells L-1). Marrouchi *et al.* (2009) also noted that the highest GYMs values in November and December of each year following a flowering of *K. selliformis* were generally detected after strong sunshine, which is also consistent with the high GYMs values detected during the same period in this study. Feki *et al.* (2003) found a positive correlation between *K. selliformis* and nitrates, and a negative one with total phosphorus, while in our study, the highest GYMs concentrations were measured in oysters from Tripoli, although the highest GYMs concentrations were measured in the *Patella rustica* mussel complex in samples from Tyre, where the highest nitrates, nitrites, and chlorophyll-a concentrations and the lowest phosphates were also measured (Fig. **5**; Table **2**). A longer data set is needed to better understand the correlation between measured GYM concentrations and physico-chemical parameters.

SPXs are produced by the dinoflagellates *Alexandrium ostenfeldii* (Cembella *et al.*, 2000; Touzet *et al.*, 2008) and *A. peruvia*num (Moestrup *et al.*, 2011; EFSA, 2010c) and represent the largest CI group (Chatzianastasiou *et al.*, 2011). *Alexandrium spp.* had been detected almost along the entire Lebanese coastal area reaching a maximum of 4752 cells L-1 in the spring period (Abboud-Abi Saab and Hassoun, 2017). During the present study, this genus was also found in three areas with the highest abundance in Tyre, which coincided with the high SPXs values measured in oysters from the same area (Figs. **7** and **8c**). As *Alexandrium spp.* are globally distributed and occur in coastal, shelf, and slope waters of the Northern and Southern Hemispheres (Taylor *et al.*, 1995; Lilly *et al.*, 2007), including the Mediterranean Sea, where the diversity of *Alexandrium spp.* appears to be higher than elsewhere due to the level of taxonomic study rather than actual distribution (Anderson *et al.*, 2012). This high diversity makes differentiation among *Alexandrium spp.* species in the present samples quite problematic.

What Do Not We Know?

Incidentally, the genus Pseudonitzschia includes more than 30 taxa, of which 11-12 are potential DA producers (Lundholm, 2011), and many other species need to be better studied to evaluate their biology, ecology, and toxicity to improve monitoring and prediction of toxic species blooms. This means that future systematic studies and improved taxonomic equipment should be used together with molecular tools to distinguish between toxic and non-toxic species in Lebanese waters. Moreover, the genus Gymnodinium includes about 234 identifiable species (Thessen *et al.*, 2012), and the light microscope is not sufficient to distinguish between Gymnodiunium species on one hand and between Gymnodinium and Karenia species on the other hand, especially because they are similar under the light microscope. Moreover, other studies have shown that GYMs are also produced by *Alexandrium ostenfeldii* (Harju *et al.*, 2016; Van de Waal *et al.*, 2015; Van Wagoner *et al.*, 2011) and *Alexandrium peruvia*num (Van Wagoner *et al.*, 2011), species that are not easily recognized under the light microscope. In addition, many studies (Medhioub *et al.*, 2009; Harju *et al.*, 2016; Busch *et al.*, 2016) have found a lack of coherence between the presence of GYMs and their potential producers, which can be attributed to the presence of undescribed species as alternative sources of the detected toxins and points out the necessity to improve morphological identification tools. In addition, the genus Alexandrium includes more than 30 morphologically defined species, at least half of which are known to be toxic or otherwise harmful (Anderson *et al.*, 2012). Under the light microscope, *Alexandrium ostenfeldii* and *A. peruvia*num are morphologically very similar but can be distinguished by their cell size, the shape of their plates, and the right anterior margin of their plates (Balech, 1995). In the genus Alexandrium, the advent of molecular techniques has challenged the classification of species based on morphological characters by showing that (i) there is a high degree of genetic diversity within the same morphospecies, and (ii) some characters used to distinguish closely related morphospecies exhibit high variability and are not consistent with molecular genetic clustering. However, these features are too difficult to distinguish with the light microscope, which is the main obstacle to further identification of *Alexandrium spp.* in Lebanese waters.

Toxicity and Regulations for the Biotoxins Detected

The European Food Safety Authority (EFSA) estimated the lowest observed adverse effect level (LOAEL) for DA neurotoxicity in humans to be 0.9 µg g-1 and established an acute reference dose (ARfD) of 30 µg kg-1 body weight (EFSA, 2009). All *S. spinosus* samples collected in Tyre had levels below the EFSA-estimated LOAEL, while 82% of *S. spinosus* collected in Tripoli had DA concentrations above the LOAEL but below the ARfD (Fig. **3**). These concentrations raise concerns about possible effects on humans as, in the short term, intoxication with DA can cause gastrointestinal symptoms within 24 hours of consuming contaminated seafood (Vilariño *et al.*, 2018). In the long term, DA can cause morphological changes in the hippocampus of rats (Pulido, 2008), and a chronic DA toxicosis syndrome has also been found in sea lions in their natural habitat (Goldstein *et al.*, 2008). These studies show that chronic effects were more common in juveniles, indicating a higher susceptibility of the developing brain to DA exposure (Doucette and Tasker, 2016). In humans, studies of Native Americans who consumed more than 15 razor clams per month with less than 20 μg DA g-1 over a four-year period show that they experienced mild memory loss (Grattan *et al.*, 2016a; Grattan *et al.*, 2016b). A recent human health study also found that high-consumption consumers of razor clams (with harmless levels of DA) had poorer everyday memory than non- or low-consumption consumers, based on dietary exposure 10 days and 1 year prior to assessment (Grattan *et al.*, 2018; more on DA symptoms in Vilariño *et al.*, 2018). This means that both short- and long-term DA neurotoxicity due to repeated consumption may be associated with low-level, chronic exposures in adults who are heavy consumers of contaminated seafood (Petroff *et al.*, 2021).

Although acute toxicity of CIs have been demonstrated by MBA-positive samples containing only gymnodimine and spirolide toxins (Ujević *et al.*, 2015), no acute poisonings in humans have been directly associated with seafood contamination (Marrouchi *et al.*, 2013; Harju *et al.*, 2016; Visciano *et al.*, 2016). There is limited information on the absorption, distribution, metabolism, and excretion of CIs in animals or humans (Chatzianastasiou *et al.*, 2011). Therefore, neither an acute reference dose (ARfD) nor a tolerable daily intake (TDI) has been proposed to prevent acute and chronic toxicity, respectively (Ben Naila *et al.*, 2012). Despite this, the chronic toxicity of GYMs remains unclear, as their role in the development of neurodegenerative diseases such as Alzheimer's or Parkinson's is controversial (Alonso *et al.*, 2011; Marques *et al.*, 2014). As a result, levels of CIs in seafood have not been regulated in the European Union or elsewhere. Gymnodimine is highly toxic when injected intraperitoneally (lowest lethal dose: LD50= 96 μg kg-1), but when ingested in the diet (> 7500 μg kg-1 in mice), GYMs show low toxicity (Munday *et al.*, 2004). In the present study, the maximum GYMs concentration was 102.9 μg kg-1, slightly higher than the LD50 value (causing acute toxicity by intraperitoneal injection in mice) and much lower than the concentration showing toxicity in mice when ingested with food (Fig. **5**). These results are consistent with the studies mentioned above, which suggest that GYMs pose a low risk to humans consuming contaminated shellfish. In addition, EFSA has classified the risk for spirolides as low, considering the 95th percentile SPXs concentrations of 9, 15, and 7 μg kg-1 in shellfish meat for mussels, oysters, and clams, respectively (Ben Naila *et al.*, 2012), which is within the concentration range recorded for the spiny oysters from the present study (Fig. 7). Furthermore, for the risk characterization of exposure to these toxins (Rambla-Alegre *et al.*, 2018), a margin of exposure approach (MOE) was used over the very limited toxicity data by dividing the LD50 value by the estimated 95th percentile of exposure from shellfish consumption. Considering that the LD50 after administration of SPXs in the diet is ~ 500 μg kg-1 body weight and the MOE is estimated to be in the range of 2941-4545 μg kg-1 (Rambla-Alegre *et al.*, 2018), it is unlikely that there is a health risk due to exposure to SPXs from seafood consumption in Lebanon.

Contamination in Non-shellfish Species

DA is not limited to mussels, oysters, and scallops; squid, sardines, anchovies, crabs, and lobsters, as other marine organisms may also contain DA (Lopez-Rivera *et al.*, 2009). Many studies have found significant DA levels in non-shellfish species such as the cephalopod common cuttlefish *Sepia officinalis* in Portugal (Costa *et al.*, 2005) and Morocco (Ben Haddouch *et al.*, 2016). These studies showed that high DA concentrations were measured in muscles (0.7 and 16 μg g−1 in Portugal and Morocco, respectively), and high DA levels were measured in the digestive glands of cuttlefish (50 and 241.7 μg g-1 in Morocco and Portugal, respectively). In the present study, DA levels in the flesh of the fish *S. rivulatus* were below the detection limit (Fig. **3**), but no measurements were made in its digestive glands. This step is imperative in future studies because in some countries, including Lebanon, whole juvenile fish are consumed (*i.e.*, without evisceration), in which case they could pose a risk to human health because DA can reach harmful levels in the digestive glands. Measurements of DA in seawater may be useful in a systematic monitoring program to prevent bioaccumulation of this toxin by marine biota, as Dursun *et al.* (2016) recorded DA in coastal waters of the Sea of Marmara in Turkey (0.96 and 5.25 μg mL-1). In addition, gymnodimines have been detected in many bivalve species, including green-shelled mussels, blue mussels, scallops, cockles, clams, oysters, and abalone (MacKenzie *et al.*, 2002; Stirling, 2001). However, other organisms such as crustaceans, gastropods, and fish have also been reported as vectors (Shumway, 1995; Deeds *et al.*, 2008), including the gastropod *Phorcus turbinatus* and the fish *S. rivulatus*, as demonstrated in this study.

Biotoxins and Climate Change

The frequency and severity of HABs are predicted to increase as climate change progresses (Gobler, 2020), partly due to changes associated with eutrophication (Anderson *et al.*, 2002; Rabalais *et al.*, 2009; Kudela *et al.*, 2010) and warming (O'Neil *et al.* 2012; Paerl *et al.* 2016; Ryan *et al.* 2017), as well as other stressors (*e.g.*, ocean acidification, deoxygenation, *etc.*). Globally, several studies have highlighted the impact of climate change on seafood safety hazards at different stages of the food chain, from primary production to consumption (Tirado *et al.*, 2010; Morabito *et al.*, 2017). Anthropogenic pressures such as coastal urbanization, eutrophication, ballast water discharge, and other stressors are already impacting marine life, including phytoplankton populations, creating suitable conditions for HABs (Stachowich *et al.*, 2002; Hallegraeff, 2010; Picot *et al.*, 2011; Valbi *et al.*, 2019).

The Mediterranean Sea is changing, so is its marine biodiversity, especially toxic phytoplankton species, which have increased dramatically in the last decade with 15 new species reported each year, mainly in the Levantine Sea (Garcés *et al.*, 2000; EEA, 2006; Aligizaki *et al.*, 2009; Faimali *et al.*, 2012; Templado, 2014; Vlamis and Katikou, 2015). An example is *Ostreopsis ovata*, a tropical species that has recently bloomed in both the Western and Eastern Mediterranean basins, causing the death of benthic organisms and several human health problems (Mangialajo *et al.*, 2011; Amzil *et al.*, 2012). Some toxic *dinoflagellates* that cause HABs also benefit from the effects of climate change on land runoff, terrestrial inputs, nutrient availability, water mass circulation, and water column stratification (Goffart *et al.*, 2002; Katikou, 2020), such as *Alexandrium minutum* in the Mediterranean Sea (Valbi *et al.*, 2019). For example, temperature likely supports the spread of K. selliformis (Feki *et al.*, 2003), but also this species had a positive relationship with nitrate and a negative relationship with total phosphorus. The different spatial gradients of these two chemical variables resulted in spatial differences in the development of *K. selliformis*, which was prevalent near tourist areas (Feki *et al.*, 2003). In general, HABs increase in parallel with climate change (Griffith and Gobler, 2019). Therefore, conducting studies that investigate the interactive effects between climate change stressors, HAB species, and biotoxins in a climate change hotspot such as the Mediterranean Sea, would be interesting. In Lebanon, phytoplankton species are studied in a long-term context, but this is the first study reporting biotoxin levels in this country.

In addition, climate change and its consequences are affecting the occurrence of marine biotoxins in the marine environment, for example, by exacerbating the presence of biotoxins and increasing the populations of pathogenic microorganisms (Marques *et al.*, 2010). Climate change affects phytoplankton structure and diversity through changes in temperature, salinity, precipitation, nutrients, and dissolved oxygen; increased frequency and intensity of extreme weather events; ocean warming and acidification; changes in transport pathways for contaminants; and invasive species invasions. All of these new patterns affect the structure of HABs (Molgó *et al.*, 2014) and have been linked to the increase in new algal phenomena (Hallegraeff, 2010). Therefore, a better understanding of the factors controlling HABs is urgently needed to define and detect bloom events and quantify the important physical and biogeochemical factors that create conditions for increased algal growth and/or biotoxins' accumulation.

CONCLUSION AND RECOMMENDATIONS

In this study, the presence of lipophilic toxins and cyclic imines in marine biota from different areas of the Lebanese coast was demonstrated for the first time.

- Values below the detection limit (LOD) were determined for okadaic acid (OA), dinophysistoxin-1 and 2 (DTX-1,2), pectenotoxin-1 and 2 (PTX-1,2), yessotoxins (YTXs), and azaspiracids (AZAs). Only domoic acid (DA), gymnodimine (GYMb), and spirolide (SPX) were found above the detection limit in some species/areas.
- The results show that the highest levels of domoic acid and gymnodimine (3.88 mg DA kg-1 and 102.9 μg GYM kg-1, respectively) were found in spiny oysters (*S. spinosus*) collected in Tripoli, in agreement with the presence of large amounts of genera producing these toxins in the same station (Pseudo-nitzchia spp. and Gymndinium spp.), while the highest concentrations of spirolides (15.07 μg SPX kg-1) were also found in spiny oysters but from Tyre, consistent with the high abundance of producers of these toxins (*Alexandrium spp.*) in this area.
- Concentrations of DA were below EU limits but above the Lowest Observed Adverse Effect Level (0.9 μg g-1) for neurotoxicity in humans and below the Acute Reference Dose (30 μg kg-1 body weight), both established by EFSA. These concentrations raise concerns about potential effects that could occur in humans from repeated consumption of DA-contaminated seafood.
- Considering the lowest lethal dose (LD50) after the administration of GYMs and SPXs in the mouse bioassay, it is unlikely that there is a health risk due to exposure to both toxins from seafood consumption in Lebanon. Nevertheless, the chronic toxicity of GYMs and SPXs remains unclear, and further research is needed to establish tolerable daily intake (TDI) limits for human consumption.
- The results shed light on the importance of the oyster (spiny oyster: *Spondylus spinosus*) as a sentinel species that biologically integrates various toxins. Since biotoxins have been found in species other than bivalves, it is important to monitor phycotoxins not only in shellfish and commercial species, but also in various organisms in the marine food chain, as they can act as vectors for toxins in commercial/edible seafood and eventually into humans.

The main obstacle to a better identification of potential biotoxin producers in Lebanese waters is the difficult taxonomic aspects of phytoplankton species in general. As Taylor (1985) has noted, "Nowhere is the value of taxonomy more evident than in its application to toxic species". Traditional criteria and tools (such as the light microscope) used to identify phytoplankton species may not be as helpful in the taxonomy of toxic species because the "morphospecies" are different and other criteria must be used in addition to purely morphological criteria (Taylor, 1991; Cembella, 2003). Therefore, new developments in identification (*e.g.*, genetic and molecular tools, electronic microscopes, flow cytometers, *etc.*) have reached a stage where they are routinely used in research and some monitoring programs in recent decades (Anderson *et al.*, 2005b). These new tools can help better identify new potential biotoxin producers in Lebanese waters to upgrade the monitoring system in that few highly populated areas.

Furthermore, filter-feeding shellfish do not require a dense bloom of toxic algae to eventually accumulate toxin levels harmful to humans, as many of the most serious algal-related hazards to human health are not necessarily associated with a dense, obvious bloom (Wu *et al.*, 2015). Because the occurrence of high phycotoxins does not always coincide with a bloom of the potential producer (as discussed in one of a previous section), there may be new, and as yet undescribed species, as alternative sources of the detected toxins, and better identification tools are also needed in this case. In addition, most official monitoring programs analyze raw molluscan tissue to determine levels of lipophilic toxins. However, most shellfish are consumed cooked or steamed, so heat-treated tissue seems to be the most relevant form for analysis. In addition, heat treatment may result in a two-fold concentration of toxins due to water loss. These aspects should be considered in future risk assessments to establish and control acceptable toxin levels (Wu *et al.*, 2015).

Although in our study, the levels of DA were below the prescribed limits and the levels of cyclic imines are not yet regulated, it is crucial to set limits for the consumption of marine organisms, and it has become a concern to assess the short- and long-term risks to human health, as their repeated consumption, could be hazardous to consumers. In this regard, a consensus is emerging that further studies should be conducted to improve our understanding of the potential producers, their ecology and distribution, their ecotoxicological behavior, and their toxicity in marine organisms and humans (*i.e.*, gastro-intestinal absorption, tissue disposition, and crossing of the blood-brain and placental barriers, *etc.*). Considerable progress has been made for characterizing the toxin-producing phytoplankton genera, but the genes involved in their production and the pathways leading to the biosynthesis of the various toxin families remain to be more thoroughly investigated. In addition, the ecological factors that favor HABs need to be better determined and delimited, and more information is needed on the environmental distribution and risks of chronic exposure to these phycotoxins (Picot *et al.*, 2011; Molgó *et al.*, 2017; Farabegoli *et al.*, 2018).

Finally, a regular monitoring program is required to establish a reliable, accurate estimate of bloom toxicity. Such adequate and systematic monitoring efforts will help address the potential impacts of marine bio-toxins on marine species, commercial and non-commercial, as well as on human health.

ACKNOWLEDGEMENTS

The authors would like to thank Mr. Elie Tarek, Mr. Houssein Jaber, Mr. Salim Kabbara for their assistance during the sampling and preparation of samples in Lebanon, Ms. Abeer Ghanem and Ms. Marie-Thérèse Kassab for their assistance in the measurements of nutrients and chlorophyll-a concentrations. This study was supported through the EU-funded project ELME (Evaluation of the Lebanese Marine Environment: a multidisciplinary study)/Reference number: ENI/2018/398-295. Also, the study was funded by the Ministry of Science and Education of the Republic of Croatia as a part of Multiannual Financing intended for institutions.

REFERENCES

Abboud-Abi Saab, M., Chedid, S., Kassab, M-T. (2006). The effect of environmental factors on the development of potentially harmful microalgae in fishing harbors in the Lebanese waters (Eastern Mediterranean). In: Chouikhi, A., Kouyoumjian, H.H, (Eds.), *Protection of coastal & marine environment* Inter-Islami Science and Technology Network on Oceanography (INOC).(pp. 9-11). Izmir-Turkey:

Abboud-Abi Saab, M., Fakhri, M., Sadek, E., Mattar, N. (2008). An estimate of the environmental status of Lebanese littoral waters using nutrients and Chlorophyll a as indicators. *Leban. Sci. J., 9*(1), 43-60.

Abboud-Abi Saab, M., Fakhri, M., Kassab, M-T., Mattar, N. (2008). Phénomène exceptionnel d'eaux colorées au printemps 2007 dans la zone côtière libanaise de Zouk-Nahr El Kelb. *Leban. Sci. J., 9*(1), 61-70. b

Alarcan, J., Biré, R., Le Hégarat, L., Fessard, V. (2018). Mixtures of lipophilic phycotoxins: Exposure data and toxicological assessment. *Mar. Drugs, 16*(2), 46.
[http://dx.doi.org/10.3390/md16020046] [PMID: 29385038]

Aligizaki, K., Katikou, P., Nikolaidis, G. (2009). Toxic benthic dinoflagellates and potential risk in the mediterranean Sea. *Int. Conf. Mollu.Shellf. Saf.*Nantes

Alonso, E., Vale, C., Vieytes, M.R., Laferla, F.M., Giménez-Llort, L., Botana, L.M. (2011). The cholinergic antagonist gymnodimine improves Aβ and tau neuropathology in an *in vitro* model of Alzheimer disease. *Cell. Physiol. Biochem., 27*(6), 783-794.
[http://dx.doi.org/10.1159/000330086] [PMID: 21691095]

Amer, L.A., Benali, I., Dermeche, S., Bouderbala, M. Seasonal variations of the biometric indices of Patella rustica Linnaeus,1758 (*Gastropoda Patellidae*) from contrasted sites of the westerm Algerian coast. Biodivers. J., 9, 205-212.

Amzil, Z., Fresnel, J., Le Gal, D., Billard, C. (2001). Domoic acid accumulation in French shellfish in relation to toxic species of *Pseudo-nitzschia* multiseries and *P. pseudodelicatissima. Toxicon, 39*(8), 1245-1251.
[http://dx.doi.org/10.1016/S0041-0101(01)00096-4] [PMID: 11306137]

Amzil, Z., Sibat, M., Royer, F., Masson, N., Abadie, E. (2007). Report on the first detection of pectenotoxin-2, spirolide-a and their derivatives in French shellfish. *Mar. Drugs, 5*(4), 168-179.
[http://dx.doi.org/10.3390/md504168] [PMID: 18463732]

Amzil, Z., Sibat, M., Chomerat, N., Grossel, H., Marco-Miralles, F., Lemee, R., Nezan, E., Sechet, V. (2012). Ovatoxin-a and palytoxin accumulation in seafood in relation to Ostreopsis cf. ovata blooms on the French

Mediterranean coast. *Mar. Drugs, 10*(12), 477-496.
[http://dx.doi.org/10.3390/md10020477] [PMID: 22412814]

Anderson, D.M. (1994). Red Tides. *Sci. Am., 271*(2), 62-68.
[http://dx.doi.org/10.1038/scientificamerican0894-62] [PMID: 8066432]

Anderson, D.M. (1997). Turning back the harmful red tide. *Nature, 388*(6642), 513-514.
[http://dx.doi.org/10.1038/41415]

Anderson, D.M., Kulis, D.M., Keafer, B.A., Gribble, K.E., Marin, R., Scholin, C.A. (2005). Identification and enumeration of *Alexandrium spp.* from the Gulf of Maine using molecular probes. *Deep Sea Res. Part II Top. Stud. Oceanogr., 52*(19-21), 2467-2490.
[http://dx.doi.org/10.1016/j.dsr2.2005.06.015]

Anderson, D.M., Fensin, E., Gobler, C.J., Hoeglund, A.E., Hubbard, K.A., Kulis, D.M., Landsberg, J.H., Lefebvre, K.A., Provoost, P., Richlen, M.L., Smith, J.L., Solow, A.R., Trainer, V.L. (2021). Marine harmful algal blooms (HABs) in the United States: History, current status and future trends. *Harmful Algae, 102*, 101975.
[http://dx.doi.org/10.1016/j.hal.2021.101975] [PMID: 33875183]

Arapov, J., Ivana, U., Živana Ninčević, G., Sanda, S., Alfiero, C., Anna, M., Silvia, P., Elena, P., Vilar-González, A., Maria Luisa, R.V, Nikša, N., Ivona, M. (2015). Shellfish lipophilic toxin profile and toxic phytoplankton species along eastern adriatic coast. *Fresen. Environ. Bulle., 24*(12).

Aune, T., Espenes, A., Aasen, J.A.B., Quilliam, M.A., Hess, P., Larsen, S. (2012). Study of possible combined toxic effects of azaspiracid-1 and okadaic acid in mice *via* the oral route. *Toxicon, 60*(5), 895-906.
[http://dx.doi.org/10.1016/j.toxicon.2012.06.007] [PMID: 22750012]

Balech, E. (1995). *The genus alexandrium halim (dinoflagellata)..* Cork, Ireland: Sherkin Island Marine Station, Sherkin Island, Co.

Bates, S.S., Katherine, A., Nina, L., Marina, M., Chui Pin, L. (2018). Pseudo-nitzschia, Nitzschia, and domoic acid: New research since 201. *Harmful Algae, 79*, 3-43.
[http://dx.doi.org/10.1016/j.hal.2018.06.001] [PMID: 30420013]

Belin, C. (1993). Distribution of *dinophysis spp.* and *alexandrium minutum* along french coasts since 1984, and their DSP and PSP toxicity levels. In: Smayda, T.J., Shimizu, Y., (Eds.), *Toxic phytoplankton blooms in the sea.* (pp. 469-474). Amsterdam: Elsevier Science Publishers.

Ben haddouch, A., Hamid, T., Hind, E., Samir Ben, B., Btissam, E., Khalid, M., Abdeltif, B., Fatima, M., Asmaa, R., Loutfi, M. (2016). Accumulation and tissue distribution of domoic acid in the common cuttlefish, sépia officinalis from the south moroccan coast. *Amer. Scient.Res. J. Eng. Techn. Sci. (ASRJETS), 15*(1).

Naila, I.B., Hamza, A., Gdoura, R., Diogène, J., de la Iglesia, P. (2012). Prevalence and persistence of gymnodimines in clams from the Gulf of Gabes (Tunisia) studied by mouse bioassay and LC–MS/MS. *Harmful Algae, 18*, 56-64.
[http://dx.doi.org/10.1016/j.hal.2012.04.004]

Berdalet, E., Fleming, L.E., Gowen, R., Davidson, K., Hess, P., Backer, L.C., Moore, S.K., Hoagland, P., Enevoldsen, H. (2015). Marine harmful algal blooms, human health and wellbeing: challenges and opportunities in the 21st century. *J. Mar. Biol. Assoc. U. K., 2015*, 61-91.
[PMID: 26692586]

Bergasa, O., Ramírez, R., Collado, C., Hernández-Brito, J.J., Gelado-Caballero, M.D., Rodríguez-Somozas, M., Haroun, R.J. (2007). Study of metals concentration levels in Patella piperata throughout the Canary Islands, Spain. *Environ. Monit. Assess., 127*(1-3), 127-133.
[http://dx.doi.org/10.1007/s10661-006-9266-x] [PMID: 17171292]

Bjørnebekk, A., Westlye, L.T., Walhovd, K.B., Fjell, A.M. (2010). Everyday memory: Self-perception and structural brain correlates in a healthy elderly population. *J. Int. Neuropsychol. Soc., 16*(6), 1115-1126.
[http://dx.doi.org/10.1017/S1355617710001025] [PMID: 20946708]

Biré, R., Krys, S., Frémy, J.M., Dragacci, S., Stirling, D., Kharrat, R. (2002). First evidence on occurrence of

gymnodimine in clams from Tunisia. *J. Nat. Toxins, 11*(4), 269-275.
[PMID: 12503869]

Bouchouicha-Smida, D. (2015). Detection of domoic acid in *Mytilus galloprovincialis* and *Ostrea edulis* linked to the presence of *Nitzschia bizertensis* in Bizerte Lagoon (SW Mediterranean). *Estu. Coast. Shelf. Sci., 165*(5), 270-278.
[http://dx.doi.org/10.1016/j.ecss.2015.05.029]

Busch, J.A., Andree, K.B., Diogène, J., Fernández-Tejedor, M., Toebe, K., John, U., Krock, B., Tillmann, U., Cembella, A.D. (2016). Toxigenic algae and associated phycotoxins in two coastal embayments in the Ebro Delta (NW Mediterranean). *Harmful Algae, 55*, 191-201.
[http://dx.doi.org/10.1016/j.hal.2016.02.012] [PMID: 28073532]

Cembella, A.D., Lewis, N.I., Quilliam, M.A. (2000). The marine dinoflagellate *Alexandrium ostenfeldii* (dinophyceae) as the causative organism of spirolide shellfish toxins. *Phycologia, 39*(1), 67-74.
[http://dx.doi.org/10.2216/i0031-8884-39-1-67.1]

Cembella, A.D. (2003). Chemical ecology of eukaryotic microalgae in marine ecosystems. *Phycologia, 42*(4), 420-447.
[http://dx.doi.org/10.2216/i0031-8884-42-4-420.1]

Ciminiello, P., Dell'Aversano, C., Fattorusso, E., Forino, M., Magno, G.S., Tartaglione, L., Quilliam, M.A., Tubaro, A., Poletti, R. (2005). Hydrophilic interaction liquid chromatography/mass spectrometry for determination of domoic acid in Adriatic shellfish. *Rapid Commun. Mass Spectrom., 19*(14), 2030-2038.
[http://dx.doi.org/10.1002/rcm.2021] [PMID: 15962348]

Cloern, J.E., Dufford, R. (2005). Phytoplankton community ecology: principles applied in San Francisco Bay. *Mar. Ecol. Prog. Ser., 285*, 11-28.
[http://dx.doi.org/10.3354/meps285011]

Congestri, R., Polizzano, S., Albertano, P. (2008). Toxic pseudo-nitzschia populations from the middle tyrrhenian sea (mediterranean sea, Italy). In: Evangelista, V., Barsanti, L., Frassanito, A.M., Passarelli, V., Gualtieri, P., (Eds.), *Algal Toxins: Nature, Occurrence, Effect and Detection. NATO Science for Peace and Security Series A: Chemistry and Biology.*. Dordrecht: Springer.
[http://dx.doi.org/10.1007/978-1-4020-8480-5_7]

Corriere, M., Soliño, L., Costa, P.R. (2021). Effects of the marine biotoxins okadaic acid and dinophysistoxins on fish. *J. Mar. Sci. Eng., 9*(3), 293.
[http://dx.doi.org/10.3390/jmse9030293]

Costa, P.R., Rosa, R., Duarte-Silva, A., Brotas, V., Sampayo, M.A.M. (2005). Accumulation, transformation and tissue distribution of domoic acid, the amnesic shellfish poisoning toxin, in the common cuttlefish, Sepia officinalis. *Aquat. Toxicol., 74*(1), 82-91.
[http://dx.doi.org/10.1016/j.aquatox.2005.01.011] [PMID: 15961171]

Ćustović, S., Orhanović, S., Ninčević-Gladan, Ž., Milandri, A., Pavela-Vrančić, M. (2012). The presence of yessotoxins and gymnodimine in the mussel *mytilus galloprovincialis* from the southern part of adriatic sea (Dubrovnik area, Croatia). *Fresenius Environ. Bull., 21*, 3842-3846.

Deeds, J., Landsberg, J., Etheridge, S., Pitcher, G., Longan, S. (2008). Non-traditional vectors for paralytic shellfish poisoning. *Mar. Drugs, 6*(2), 308-348.
[http://dx.doi.org/10.3390/md6020308] [PMID: 18728730]

Diogène, J. (2017). Marine toxin analysis for the benefit of 'one health' and for the advancement of science. *Compr. Anal. Chem., 78*, 1-34.
[http://dx.doi.org/10.1016/bs.coac.2017.08.001]

Donald, K.M., Preston, J., Williams, S.T., Reid, D.G., Winter, D., Alvarez, R., Buge, B., Hawkins, S.J., Templado, J., Spencer, H.G. (2012). Phylogenetic relationships elucidate colonization patterns in the intertidal grazers Osilinus Philippi, 1847 and Phorcus Risso, 1826 (Gastropoda: Trochidae) in the northeastern Atlantic Ocean and Mediterranean Sea. *Mol. Phylogenet. Evol., 62*(1), 35-45.

[http://dx.doi.org/10.1016/j.ympev.2011.09.002] [PMID: 21945534]

Doucette, T.A., Tasker, R.A. (2015). Perinatal domoic acid as a neuroteratogen. *Curr. Top. Behav. Neurosci., 29*, 87-110.
[http://dx.doi.org/10.1007/7854_2015_417] [PMID: 26695171]

Draisci, R., Lucentini, L., Giannetti, L., Boria, P., Poletti, R. (1996). First report of pectenotoxin-2 (PTX-2) in algae (*Dinophysis fortii*) related to seafood poisoning in Europe. *Toxicon, 34*(8), 923-935.
[http://dx.doi.org/10.1016/0041-0101(96)00030-X] [PMID: 8875779]

Durborow, R.M. (1999). Health and safety concerns in fisheries and aquaculture. *Occup. Med., 14*(2), 373-406.
[PMID: 10329911]

Dursun, F., Yurdun, T., Ünlü, S. (2016). The first observation of domoic acid in plankton net samples from the sea of marmara, Turkey. *Bull. Environ. Contam. Toxicol., 96*(1), 70-75.
[http://dx.doi.org/10.1007/s00128-015-1704-4] [PMID: 26615530]

EEA. (2006). Priority issues in the Mediterranean environment. Report no 4/2006 European Enviroment Agency. Copenhagen.

EFSA. (2008). European Food Safety Authority, Scientific Opinion: Marine Biotoxins in Shellfish – Okadaic acid and analogues. *EFSA, 589*, 1-62.

EFSA. (2009). Scientific Opinion of the Panel on Contaminants in the Food Chain on a request from the European Commission on Marine biotoxins in shellfish–domoic acid. *EFSA J., 1181*, 1-61.

EFSA. (2010). European food safety authority, scientific opinion: Marine biotoxins in shellfish : Cyclic imines. *EFSA. J., 86*, , 1628.

EFSA. (2010). Scientific Opinion on marine biotoxins in shellfish : Emerging toxins: Brevetoxin group. *EFSA. J., 8*(7), 1677.

EFSA. (2010). Scientific Opinion on marine biotoxins in shellfish – Cyclic imines (spirolides, gymnodimines, pinnatoxins and pteriatoxins). EFSA Panel on Contaminants in the Food Chain. *CONTAM, 8*(6), 1628.
[http://dx.doi.org/10.2903/j.efsa.2010.1628]

Ettoubi, E., Akif, F., Genten, F., Blaghen, M. (2020). Marine Biotoxins: Origins, effects, distribution, prevention and treatment. *Int. J. Innov. Sci. Res. Tech., 5*(11).

EU-NRL. (2001). Minutes of the 4th meeting of EU National Reference Laboratories (EU-NRL) on marine biotoxins. 29-31 October 2001. Vigo, Spain.

EU-NRL. (2002). Minutes of 5th meeting of EU National Reference Laboratories (EU-NRL) on marine biotoxins. 10-12 December 2002. Brussels, Belgium.

European Reference Laboratory for Marine Biotoxins (EURLMB). (2015). EU Harmonised standard operating procedure for determination of lipophilic marine biotoxins in molluscs by LC-MS/MS. Version 5.

Faimali, M., Giussani, V., Piazza, V., Garaventa, F., Corrà, C., Asnaghi, V., Privitera, D., Gallus, L., Cattaneo-Vietti, R., Mangialajo, L., Chiantore, M. (2012). Toxic effects of harmful benthic dinoflagellate Ostreopsis ovata on invertebrate and vertebrate marine organisms. *Mar. Environ. Res., 76*, 97-107.
[http://dx.doi.org/10.1016/j.marenvres.2011.09.010] [PMID: 22000703]

FAO/IOC/WHO. (2004). Joint FAO/IOC/WHO expert meeting.

FAO/IOC/WHO. (2006). Food Standards Programme. *Proceedings of the Codex Committee on Fish and Fishery Products, Twenty-Eighth Session; Beijing, China.*

Farabegoli, F., Blanco, L., Rodríguez, L., Vieites, J., Cabado, A. (2018). Phycotoxins in marine shellfish: Origin, Occurrence and Effects on Humans. *Mar. Drugs, 16*(6), 188.
[http://dx.doi.org/10.3390/md16060188] [PMID: 29844286]

Feki, W., Hamza, A., Frossard, V., Abdennadher, M., Hannachi, I., Jacquot, M., Belhassen, M., Aleya, L.

(2013). What are the potential drivers of blooms of the toxic dinoflagellate *Karenia selliformis*? A 10-year study in the Gulf of Gabes, Tunisia, southwestern Mediterranean Sea. *Harmful Algae, 23*, 8-18. [http://dx.doi.org/10.1016/j.hal.2012.12.001]

Fenchel, T. (1988). Marine plankton food chains. *Annu. Rev. Ecol. Syst., 19*(1), 19-38. [http://dx.doi.org/10.1146/annurev.es.19.110188.000315]

Ferrante, M, Conti, GO, Ledda, C, Zuccarello, M (2009). First results about an ostreopsis ovata monitoring along the catania coast (Sicily-Italy). *Epidemiology, 20*(6), S159.

Ferriss, B.E., Marcinek, D.J., Ayres, D., Borchert, J., Lefebvre, K.A. (2017). Acute and chronic dietary exposure to domoic acid in recreational harvesters: A survey of shellfish consumption behavior. *Environ. Int., 101*, 70-79. [http://dx.doi.org/10.1016/j.envint.2017.01.006] [PMID: 28109640]

Ferron, P.J., Hogeveen, K., De Sousa, G., Rahmani, R., Dubreil, E., Fessard, V., Le Hegarat, L. (2016). Modulation of CYP3A4 activity alters the cytotoxicity of lipophilic phycotoxins in human hepatic HepaRG cells. *Toxicol. In Vitro, 33*, 136-146. [http://dx.doi.org/10.1016/j.tiv.2016.02.021] [PMID: 26956883]

Fonda Umani, S. (1996). Pelagic production and biomass in the Adriatic Sea. *Sci. Mar., 60*, 65-77.

Friedman, M., Fleming, L., Fernandez, M., Bienfang, P., Schrank, K., Dickey, R., Bottein, M.Y., Backer, L., Ayyar, R., Weisman, R., Watkins, S., Granade, R., Reich, A. (2008). Ciguatera fish poisoning: Treatment, prevention and management. *Mar. Drugs, 6*(3), 456-479. [http://dx.doi.org/10.3390/md6030456] [PMID: 19005579]

Fujiki, H., Suganuma, M. (1993). Tumor promotion by inhibitors of protein phosphatases 1 and 2A: The okadaic acid class of compounds. *Adv. Cancer Res., 61*, 143e194.

Haddouch, A.B., Amanhi, R., Amzil, Z., Taleb, H., Rovillon, G., Adly, F., Loutfi, M. (2017). Lipophilic toxin profile in *mytilus galloprovincialis* from the north atlantic coast of morocco: LC-MS/MS and mouse bioassay analyses. *Int. J. Sci.Res., 6*(2), 186.

Hallegraeff, G.M. (1993). A review of harmful algal blooms and their apparent global increase. *Phycologia, 32*(2), 79-99. [http://dx.doi.org/10.2216/i0031-8884-32-2-79.1]

Hallegraeff, G.M. (2008). Harmful algal blooms, coastal eutrophication and climate change. *Biol. Mar. Mediterr., 15*, 6-15.

Hallegraeff, G.M. (2010). Ocean climate change, phytoplankton community responses, and harmful algal blooms: A formidable predictive challenge. *J. Phycol., 46*(2), 220-235. [http://dx.doi.org/10.1111/j.1529-8817.2010.00815.x]

Harrness. (2005). *Harmful Algal Research and Response: A National Environmental Science Strategy 2005–2015.* Ramsdell, J.S., Anderson, D.M., Glibert, P.M. Ecological Society of America.(p. 96). Washington DC:

Hassoun, A.E.R., Ujević, I., Mahfouz, C., Fakhri, M., Roje-Busatto, R., Jemaa, S., Nazlić, N. (2021). Occurrence of domoic acid and cyclic imines in marine biota from Lebanon-Eastern Mediterranean Sea. *Sci. Total Environ., 755*(Pt 1), 142542. Available from: https://www.sciencedirect.com/science/article/abs/pii/S004896972036071X [http://dx.doi.org/10.1016/j.scitotenv.2020.142542] [PMID: 33035983]

Henriques, P., Delgado, J., Sousa, R. (2017). Patellid limpets: An overview of the biology and conservation of keystone species of the rocky shores. In: Ray, S., (Ed.), *Organismal and Molecular Malacology.* (1ˢᵗ ed., pp. 71-95). Croatia: InTech. [http://dx.doi.org/10.5772/67862]

Ito, E., Satake, M., Ofuji, K., Kurita, N., McMahon, T., James, K., Yasumoto, T. (2000). Multiple organ damage caused by a new toxin azaspiracid, isolated from mussels produced in Ireland. *Toxicon, 38*(7), 917-930.

[http://dx.doi.org/10.1016/S0041-0101(99)00203-2]

Gago-Martínez, A., Rodríguez-Vázquez, J.A. (2000). Marine Toxins: Chromatography. Reference module in chemistry. Molecular Sciences and Chemical Engineering. Encyclopedia of Separation Science.
[http://dx.doi.org/10.1016/B0-12-226770-2/04881-X]

Garcés, E., Masó, M., Vila, M., Camp, J. (2000). HABs events in the Mediterranean Sea: are they increasing? A case study the last decade in the NW Mediterranean and the genus Alexandrium. *Harmful Algae News, 20,* 1-11.

García, C., Oyaneder-Terrazas, J., Contreras, C., del Campo, M., Torres, R., Contreras, H.R. (2016). Determination of the toxic variability of lipophilic biotoxins in marine bivalve and gastropod tissues treated with an industrial canning process. *Food Addit. Contam. Part A Chem. Anal. Control Expo. Risk Assess., 33*(11), 1711-1727.
[http://dx.doi.org/10.1080/19440049.2016.1239032] [PMID: 27646025]

Giménez Papiol, G., Casanova, A., Fernández-Tejedor, M., de la Iglesia, P., Diogène, J. (2013). Management of domoic acid monitoring in shellfish from the Catalan coast. *Environ. Monit. Assess., 185*(8), 6653-6666.
[http://dx.doi.org/10.1007/s10661-012-3054-6] [PMID: 23275095]

Arapov, J., Ceredi, A., Milandri, A., Marasovic, I., Nincevic Gladan, Z., Pigozzi, S., Skejic, S., Ujevic, I. (2011). Lipophilic toxin profile in *Mytilus galloprovincialis* during episodes of diarrhetic shellfish poisoning (DSP) in the N.E. Adriatic Sea in 2006. *Molecules, 16*(1), 888-899.
[http://dx.doi.org/10.3390/molecules16010888] [PMID: 21258296]

Gobler, C.J., Owen, M., Theresa, L., Hattenrath, L., Andrew W., G., Yoonja, K., Wayne, L. (2017). Ocean warming has expanded niche of toxic algae. *Proc. Natl. Acad. Sci., 114*(19), 4975-4980.
[http://dx.doi.org/10.1073/pnas.1619575114] [PMID: 28439007]

Gobler, C.J. (2020). Climate change and harmful algal blooms: Insights and perspective. *Harmful Algae, 91,* 101731.
[http://dx.doi.org/10.1016/j.hal.2019.101731] [PMID: 32057341]

Goffart, A., Hecq, J.H., Legendre, L. (2002). Changes in the development of the winter-spring phytoplankton bloom in the Bay of Calvi (NW Mediterranean) over the last two decades: A response to changing climate? *Mar. Ecol. Prog. Ser., 236*, 45-60.
[http://dx.doi.org/10.3354/meps236045]

Goldstein, T., Mazet, J.A.K., Zabka, T.S., Langlois, G., Colegrove, K.M., Silver, M., Bargu, S., Van Dolah, F., Leighfield, T., Conrad, P.A., Barakos, J., Williams, D.C., Dennison, S., Haulena, M., Gulland, F.M.D. (2008). Novel symptomatology and changing epidemiology of domoic acid toxicosis in California sea lions (*Zalophus californianus*): an increasing risk to marine mammal health. *Proc. Biol. Sci., 275*(1632), 267-276.
[http://dx.doi.org/10.1098/rspb.2007.1221] [PMID: 18006409]

Grattan, L.M., Holobaugh, S., Morris, J.G., Jr (2016). Harmful algal blooms and public health. *Harmful Algae, 57*, 2-8.
[http://dx.doi.org/10.1016/j.hal.2016.05.003]

Grattan, L.M., Boushey, C., Tracy, K., Trainer, V.L., Roberts, S.M., Schluterman, N., Morris, J.G., Jr (2016). The association between razor clam consumption and memory in the CoASTAL cohort. *Harmful Algae, 57*, 20-25. b
[http://dx.doi.org/10.1016/j.hal.2016.03.011]

Grattan, L., Boushey, C., Liang, Y., Lefebvre, K., Castellon, L., Roberts, K., Toben, A., Morris, J. (2018). Repeated dietary exposure to low levels of domoic acid and problems with everyday memory: Research to public health outreach. *Toxins, 10*(3), 103.
[http://dx.doi.org/10.3390/toxins10030103] [PMID: 29495583]

Griffith, A.W., Gobler, C.J. (2019). Harmful algal blooms: A climate change co-stressor in marine and freshwater ecosystems. *Harmful Algae, 91*, 101590.
[http://dx.doi.org/10.1016/j.hal.2019.03.008] [PMID: 32057338]

Grupta, S.K., Singh, J. (2011). Evaluation of mollusc as sensitive indicator of heavy metal pollution in aquatic system: A review. *Inst. Integ. Omics. Appl. Biotech. J., 2*(1), 49-57.

Guéret, S.M., Brimble, M.A. (2010). Spiroimine shellfish poisoning (SSP) and the spirolide family of shellfish toxins: Isolation, structure, biological activity and synthesis. *Nat. Prod. Rep., 27*(9), 1350-1366.
[http://dx.doi.org/10.1039/c005400n] [PMID: 20585694]

Guiry, M.D. (2012). How many species of algae are there? *J. Phycol., 48*(5), 1057-1063.
[http://dx.doi.org/10.1111/j.1529-8817.2012.01222.x] [PMID: 27011267]

Kaniou-Grigoriadou, I., Mouratidou, T., Katikou, P. (2005). Investigation on the presence of domoic acid in greek shellfish. *Harmful Algae, 4*(4), 717-723.
[http://dx.doi.org/10.1016/j.hal.2004.10.002]

Karlson, B., Andersen, P., Arneborg, L., Cembella, A., Eikrem, W., John, U., West, J.J., Klemm, K., Kobos, J., Lehtinen, S., Lundholm, N., Mazur-Marzec, H., Naustvoll, L., Poelman, M., Provoost, P., De Rijcke, M., Suikkanen, S. (2021). Harmful algal blooms and their effects in coastal seas of Northern Europe. *Harmful Algae, 102*, 101989.
[http://dx.doi.org/10.1016/j.hal.2021.101989] [PMID: 33875185]

Karydis, M. (2009). Eutrophication assessment of coastal waters based on indicators: A literature review. *Glob. NEST J., 11*(4), 373-390.

Katikou, P., Aligizaki, K., Zacharaki, T., Iossifidis, D., Nikolaidis, G. (2010). First report on the presence of spirolides in Greek shellfish associated with the detection of the causative Alexandrium species. *Proc. 14th Int. Conf. Harmful. Algal. Blooms.,* 1-5.

Katikou, P., Aligizaki, K., Zacharaki, T., Iossifidis, D., Nikolaidis, G. First report on the presence of spirolides in greek shellfish associated with the detection of the causative alexandrium species. In: Pagou, KA, Hallegraeff, GM, (Eds.), *Proceedings of the 14th International Conference on Harmful Algal Blooms , 1–5, November 2010.* International Society for the Study of Harmful Algae and Intergovernmental Oceanographic Commission of UNESCO, Paris, France.

Katikou, P. (2020). 8 human impact in mediterranean coastal ecosystems and climate change: Emerging toxins. In: Luis, M., Botana, M., Carmen, L., Natalia, V., (Eds.), *Climate Change and Marine and Freshwater Toxins* (pp. 253-302). Berlin, Boston: De Gruyter.
[http://dx.doi.org/10.1515/9783110625738-008]

Kremp, A., Lindholm, T., Dreler, N., Erler, K., Gerdts, G., Eirtovaara, S., Leskinen, E. (2010). Bloom forming *Alexandrium ostenfeldii* (Dinophyceae) in shallow waters of the Finland Archipelago, Northern Baltic Sea. *Harmful Algae* Crete, Greece, edited by Pagou KA, Hallegraeff GM. International Society for the Study of Harmful Algae and Intergovernmental Oceanographic Commission of UNESCO.

Lilly, E.L., Kulis, D.M., Gentien, P., Anderson, D.M. (2002). Paralytic shellfish poisoning toxins in France linked to a human-introduced strain of Alexandrium catenella from the western Pacific: Evidence from DNAand toxin analysis. *J. Plankton Res., 24*(5), 443-452.
[http://dx.doi.org/10.1093/plankt/24.5.443]

Liu, Y., Yu, R.C., Kong, F.Z., Li, C., Dai, L., Chen, Z.F., Geng, H.X., Zhou, M.J. (2019). Contamination status of lipophilic marine toxins in shellfish samples from the Bohai Sea, China. *Environ. Pollut., 249*, 171-180.
[http://dx.doi.org/10.1016/j.envpol.2019.02.050] [PMID: 30884396]

López-Rivera, A., Pinto, M., Insinilla, A., Isla, B.S., Uribe, E., Alvarez, G., Lehane, M., Furey, A., James, K.J. (2009). The occurrence of domoic acid linked to a toxic diatom bloom in a new potential vector: The tunicate Pyura chilensis (piure). *Toxicon, 54*(6), 754-762.
[http://dx.doi.org/10.1016/j.toxicon.2009.05.033] [PMID: 19505493]

Lu, C.K., Lee, G.H., Huang, R., Chou, H.N. (2001). Spiro-prorocentrimine, a novel macrocyclic lactone from a benthic Prorocentrum sp. of Taiwan. *Tetrahedron Lett., 42*(9), 1713-1716.
[http://dx.doi.org/10.1016/S0040-4039(00)02331-5]

Mangialajo, L., Ganzin, N., Accoroni, S., Asnaghi, V., Blanfuné, A., Cabrini, M., Cattaneo-Vietti, R., Chavanon, F., Chiantore, M., Cohu, S., Costa, E., Fornasaro, D., Grossel, H., Marco-Miralles, F., Masó, M., Reñé, A., Rossi, A.M., Sala, M.M., Thibaut, T., Totti, C., Vila, M., Lemée, R. (2011). Trends in ostreopsis proliferation along the northern mediterranean coasts. *Toxicon, 57*(3), 408-420.
[http://dx.doi.org/10.1016/j.toxicon.2010.11.019] [PMID: 21145339]

Marques, A., Nunes, M.L., Moore, S.K., Strom, M.S. (2010). Climate change and seafood safety: Human health implications. *Food Res. Int., 43*(7), 1766-1779.
[http://dx.doi.org/10.1016/j.foodres.2010.02.010]

Marques, A., Rosa, R., Nunes, M.L. (2014). Seafood Safety and Human Health Implications. Goffredo, S., Dubinsky, Z. (2014). *The Mediterranean Sea – Its history and present challenges.* (pp. 589-603). New York Note 41: Author: London.
[http://dx.doi.org/10.1007/978-94-007-6704-1_36]

Marrouchi, R., Dziri, F., Belayouni, N., Hamza, A., Benoit, E., Molgó, J., Kharrat, R. (2010). Quantitative determination of gymnodimine-A by high performance liquid chromatography in contaminated clams from Tunisia coastline. *Mar. Biotechnol., 12*(5), 579-585.
[http://dx.doi.org/10.1007/s10126-009-9245-7] [PMID: 19997768]

Martens, H., Tillmann, U., Harju, K., Dell'Aversano, C., Tartaglione, L., Krock, B. (2017). Toxin variability estimations of 68 *Alexandrium ostenfeldii* (dinophyceae) strains from the netherlands reveal a novel abundant gymnodimine. *Microorganisms, 5*(2), 29.
[http://dx.doi.org/10.3390/microorganisms5020029] [PMID: 28587138]

Mattarozzi, M., Cavazza, A., Calfapietra, A., Cangini, M., Pigozzi, S., Bianchi, F., Careri, M. (2019). Analytical screening of marine algal toxins for seafood safety assessment in a protected Mediterranean shallow water environment. *Food Addit. Contam. Part A Chem. Anal. Control Expo. Risk Assess., 36*(4), 612-624.
[http://dx.doi.org/10.1080/19440049.2019.1581380] [PMID: 30882280]

McCarthy, M., Bane, V., García-Altares, M., van Pelt, F.N.A.M., Furey, A., O'Halloran, J. (2015). Assessment of emerging biotoxins (pinnatoxin G and spirolides) at Europe's first marine reserve: Lough Hyne. *Toxicon, 108*, 202-209.
[http://dx.doi.org/10.1016/j.toxicon.2015.10.007] [PMID: 26482934]

McKenzie, L. (2002). Production of gymnodimine by *Karenia selliformis*. (Haywood *et al*). *Harmful Algae,* 160-162.

McNamee, S.E., Medlin, L.K., Kegel, J., McCoy, G.R., Raine, R., Barra, L., Ruggiero, M.V., Kooistra, W.H.C.F., Montresor, M., Hagstrom, J., Blanco, E.P., Graneli, E., Rodríguez, F., Escalera, L., Reguera, B., Dittami, S., Edvardsen, B., Taylor, J., Lewis, J.M., Pazos, Y., Elliott, C.T., Campbell, K. (2016). Distribution, occurrence and biotoxin composition of the main shellfish toxin producing microalgae within European waters: A comparison of methods of analysis. *Harmful Algae, 55*, 112-120.
[http://dx.doi.org/10.1016/j.hal.2016.02.008] [PMID: 28073524]

Moestrup, Ø., Akselman, R., Cronberg, G., Elbraechter, M., Fraga, S., Halim, Y., Hansen, G., Hoppenrath, M., Larsen, J., Lundholm, N., Nguyen, L.N., Zingone, A. (2011). IOC-UNESCO Taxonomic Reference List of Harmful Micro Algae. Available from :http://www.marinespecies.org/hab/aphia.php?p=taxdetails &id=156548

Molgó, J., Marchot, P., Aráoz, R., Benoit, E., Iorga, B.I., Zakarian, A., Taylor, P., Bourne, Y., Servent, D. (2017). Cyclic imine toxins from dinoflagellates: A growing family of potent antagonists of the nicotinic acetylcholine receptors. *Cholinergic Mechanisms, 142*(S2), 41-51.
[http://dx.doi.org/10.1111/jnc.13995]

Morabito, S., Silvestro, S., Faggio, C. (2018). How the marine biotoxins affect human health. *Nat. Prod. Res., 32*(6), 621-631.
[http://dx.doi.org/10.1080/14786419.2017.1329734] [PMID: 28532167]

Moreiras, G., Leão, J.M., Gago-Martínez, A. (2019). Analysis of Cyclic Imines in Mussels (*Mytilus*

galloprovincialis) from Galicia (NW Spain) by LC-MS/MS. *Int. J. Environ. Res. Public Health, 17*(1), 281.
[http://dx.doi.org/10.3390/ijerph17010281] [PMID: 31906079]

Morgan, K.L., Larkin, S.L., Adams, C.M. (2009). Firm-level economic effects of HABS: A tool for business loss assessment. *Harmful Algae, 8*(2), 212-218.
[http://dx.doi.org/10.1016/j.hal.2008.05.002]

Moschandreou, K.K., Baxevanis, A.D., Katikou, P., Papaefthimiou, D., Nikolaidis, G., Abatzopoulos, T.J. (2012). Inter- and intra-specific diversity of *Pseudo-nitzschia* (Bacillariophyceae) in the northeastern Mediterranean. *Eur. J. Phycol., 47*(3), 321-339.
[http://dx.doi.org/10.1080/09670262.2012.713998]

Munday, R., Towers, N.R., Mackenzie, L., Beuzenberg, V., Holland, P.T., Miles, C.O. (2004). Acute toxicity of gymnodimine to mice. *Toxicon, 44*(2), 173-178.
[http://dx.doi.org/10.1016/j.toxicon.2004.05.017] [PMID: 15246766]

Nakhlé, K.F., Cossa, D., Khalaf, G., Beliaeff, B. (2006). *Brachidontes variabilis* and *Patella sp.* as quantitative biological indicators for cadmium, lead and mercury in the Lebanese coastal waters. *Environ. Pollut., 142*(1), 73-82.
[http://dx.doi.org/10.1016/j.envpol.2005.09.016] [PMID: 16343717]

Nezan, E., Chomerat, N. (2011). *Vulcanodinium rugosum* gen. nov. sp. nov. (Dinophyceae): A new marine dinoflagellate from the French Mediterranean coast. *Cryptogam., Algol., 32*, 3-18.
[http://dx.doi.org/10.7872/crya.v32.iss1.2011.003]

NOAA. (2017). *What is eutrophication?.* Available from: https://oceanservice.noaa.gov/facts/eutrophication.html

Nwankwegu, A.S., Li, Y., Huang, Y., Wei, J., Norgbey, E., Sarpong, L., Lai, Q., Wang, K. (2019). Harmful algal blooms under changing climate and constantly increasing anthropogenic actions: The review of management implications. *3 Biotech, 9*(12), 1-19.

Okaichi, T. (2004). *Red Tides.* Springer Science & Business Media.

Orellana, G., Van Meulebroek, L., De Rijcke, M., Janssen, C.R., Vanhaecke, L. (2017). High resolution mass spectrometry-based screening reveals lipophilic toxins in multiple trophic levels from the North Sea. *Harmful Algae, 64*, 30-41.
[http://dx.doi.org/10.1016/j.hal.2017.03.005] [PMID: 28427570]

Orhanovic, S., Nincevic, Z., Marasovic, I., Pavela-Vrancic, M. (1996). Phytoplankton toxins in the central adriatic sea. *Croat. Chem. Acta, 69*(1), 291-303.

Otero, A., Chapela, M.J., Atanassova, M., Vieites, J.M., Cabado, A.G. (2011). Cyclic imines: Chemistry and mechanism of action: A review. *Chem. Res. Toxicol., 24*(11), 1817-1829.
[http://dx.doi.org/10.1021/tx200182m] [PMID: 21739960]

Petroff, R., Hendrix, A., Shum, S., Grant, K.S., Lefebvre, K.A., Burbacher, T.M. (2021). Public health risks associated with chronic, low-level domoic acid exposure: A review of the evidence. *Pharmacol. Ther., 227*, 107865. Available from: https://www.sciencedirect.com/science/article/abs/pii/S016372582100067X
[http://dx.doi.org/10.1016/j.pharmthera.2021.107865] [PMID: 33930455]

Picot, C., Nguyen, T.A., Roudot, A.C., Parent-Massin, D. (2011). A preliminary risk assessment of human exposure to phycotoxins in shellfish: A review. *Hum. Ecol. Risk Assess., 17*(2), 328-366.
[http://dx.doi.org/10.1080/10807039.2011.552393]

Pigozzi, S., Bianchi, L., Boschetti, L., Cangini, M., Ceredi, A., Magnani, F., Milandri, A., Montanari, S., Pompei, M., Riccardi, E. (2008). First evidence of spirolide accumulation in northwestern Adriatic shellfish. *In: Proceedings of the 12th ICHA, 4–8 September 2006, ISSHA and IOC of UNESCO*Copenhagen, Denmark.

Powell, E.N., Klinck, J.M., Hofmann, E.E., Wilson-Ormond, E.A., Ellis, M.S. (1995). Modeling oyster populations. V. Declining phytoplankton stocks and the population dynamics of American oyster (*Crassostrea virginica*) populations. *Fish. Res., 24*(3), 199-222.

[http://dx.doi.org/10.1016/0165-7836(95)00370-P]

Pulido, O. (2008). Domoic acid toxicologic pathology: A review. *Mar. Drugs, 6*(2), 180-219.
[http://dx.doi.org/10.3390/md6020180] [PMID: 18728725]

Qiao, X., Saha, B. (2021). Quatifying the socio-economic impact of harmful algal blooms. Available from: https://www.researchgate.net/profile/Joao-Ferreir-132/publication/353669784_Quantifying_the_Soci-Economic_Impacts_of_Harmful_Algal_Blooms_in_Southwest_Florida_in_2018/links/61098e7c169a1a0103dad6e2/Quantifying-the-Socio-Economic-Impacts-of-Harmful-Algal-Blooms-in-Southwes--Florida-in-2018.pdf.

Quilliam, M.A., Gilgan, M.W., Pleasance, S., Defreitas, A.S.W., Douglas, D., Fritz, L., Hu, T., Marr, J.C., Smyth, C., Wright, J.L.C. (1993). *Toxic phytoplankton blooms in the sea.* (p. 547). Amsterdam: Elsevier.

Quilliam, M.A., Wright, J.L.C. (1989). The amnesic shellfish poisoning mystery. *Anal. Chem., 61*(18), 1053A-1060A.
[http://dx.doi.org/10.1021/ac00193a745] [PMID: 2802153]

Quiroga, I. (2006). *Pseudo-nitzschia* blooms in the bay of banyuls-sur-mer, northwestern mediterranean sea. *Diatom Res., 21*(1), 91-104.
[http://dx.doi.org/10.1080/0269249X.2006.9705654]

Rainbow, P.S. (1997). Ecophysiology of trace metal uptake in Crustaceans. *Estuar. Coast. Shelf Sci., 44*(2), 169-176.
[http://dx.doi.org/10.1006/ecss.1996.0208]

Rambla-Alegre, M., Miles, C.O., de la Iglesia, P., Fernandez-Tejedor, M., Jacobs, S., Sioen, I., Verbeke, W., Samdal, I.A., Sandvik, M., Barbosa, V., Tediosi, A., Madorran, E., Granby, K., Kotterman, M., Calis, T., Diogene, J. (2018). Occurrence of cyclic imines in European commercial seafood and consumers risk assessment. *Environ. Res., 161*, 392-398.
[http://dx.doi.org/10.1016/j.envres.2017.11.028] [PMID: 29197280]

Reñé, A., Satta, C.T., Garcés, E., Massana, R., Zapata, M., Anglès, S., Camp, J. (2011). Gymnodinium litoralis sp. nov. (Dinophyceae), a newly identified bloom-forming dinoflagellate from the NW Mediterranean Sea. *Harmful Algae, 12*, 11-25.
[http://dx.doi.org/10.1016/j.hal.2011.08.008]

Reverté, L., Soliño, L., Carnicer, O., Diogène, J., Campàs, M. (2014). Alternative methods for the detection of emerging marine toxins: Biosensors, biochemical assays and cell-based assays. *Mar. Drugs, 12*(12), 5719-5763.
[http://dx.doi.org/10.3390/md12125719] [PMID: 25431968]

Reyes-Prieto, A., Yoon, H.S., Bhattacharya, D. (2009). *Marine Algal Genomics and Evolution. Encyclopedia of Ocean Sciences* Academic Press.
[http://dx.doi.org/10.1016/B978-012374473-9.00779-7]

Richard, D., Arsenaulf, E., Cembella, A., Quilliam, M. (2001). Investigations into the toxicology and pharmacology of spirolides, a novel group of shellfish toxins, in harmful algal blooms 2000 In: Hallegraeff, G. M., Blackburn, S. I., Bolch, C. J., Lewis, R. J., (Eds.), *Proc. Ninth. Int. Conf. Harmful. Algal. Blooms.* Intergovernmental Oceanographic Commission of UNESCO.(p. 383390). Paris, France:

Rijat Leblad, B., Lundholm, N., Goux, D., Veron, B., Sagou, R., Taleb, H., Nhhala, H., Er-Raioui, H. (2013). *Pseudo-nitzschia peragallo* (bacillariophyceae) diversity and domoic acid accumulation in tuberculate cockles and sweet clams in m'diq bay, morocco. *Acta Bot. Croat., 72*(1), 35-47.
[http://dx.doi.org/10.2478/v10184-012-0004-x]

Romdhane, M.S., Eilertsen, H.C., Yahia, K.D., Yahia, M.N.D. (1998). Toxic dinoflagellate blooms in Tunisian lagoons: Causes and consequences for aquaculture. In: Reguera, B., Blanco, J., Fernandez, M., Wyatt, T., (Eds.), *Proceedings of the VIII International Conference on Harmful Algae,* June 1999Vigo, Spain80-83.

Sahraoui, I., Bates, S.S., Bouchouicha, D., Mabrouk, H.H., Hlaili, A.S. (2011). Toxicity of *Pseudo-nitzschia* populations from Bizerte Lagoon, Tunisia, southwest Mediterranean, and first report of domoic acid

production by *P. brasiliana. Diatom Res., 26*(3), 293-303.
[http://dx.doi.org/10.1080/0269249X.2011.597990]

Sahraoui, I. (2012). Response of potentially toxic *Pseudo-nitzschia (Bacillariophyceae)* populations and domoic acid to environmental conditions in a eutrophied, SW Mediterranean coastal lagoon (Tunisia). *Estu. Coastal. Shelf. Sci., 102-103*, 95-104.
[http://dx.doi.org/10.1016/j.ecss.2012.03.018]

Salgado, P., Riobó, P., Rodríguez, F., Franco, J.M., Bravo, I. (2015). Differences in the toxin profiles of *Alexandrium ostenfeldii* (Dinophyceae) strains isolated from different geographic origins: Evidence of paralytic toxin, spirolide, and gymnodimine. *Toxicon, 103*(1), 85-98.
[http://dx.doi.org/10.1016/j.toxicon.2015.06.015] [PMID: 26093028]

Seki, T., Satake, M., Mackenzie, L., Kaspar, H.F., Yasumoto, T. (1995). Gymnodimine, a new marine toxin of unprecedented structure isolated from New Zealand oysters and the dinoflagellate, Gymnodinium sp. *Tetrahedron Lett., 36*(39), 7093-7096.
[http://dx.doi.org/10.1016/0040-4039(95)01434-J]

Shumway, S.E. (1995). Phycotoxin-related shellfish poisoning: Bivalve molluscs are not the only vectors. *Rev. Fish. Sci., 3*(1), 1-31.
[http://dx.doi.org/10.1080/10641269509388565]

Silbergeld, E.K., Grattan, L., Oldach, D., Morris, J.G. (2000). Pfiesteria: Harmful algal blooms as indicators of human: Ecosystem interactions. *Environ. Res., 82*(2), 97-105.
[http://dx.doi.org/10.1006/enrs.1999.3987] [PMID: 10662524]

Silva, M., Barreiro, A., Rodriguez, P., Otero, P., Azevedo, J., Alfonso, A., Botana, L., Vasconcelos, V. (2013). New invertebrate vectors for PST, spirolides and okadaic acid in the North Atlantic. *Mar. Drugs, 11*(6), 1936-1960.
[http://dx.doi.org/10.3390/md11061936] [PMID: 23739043]

Morabito, S., Silvestro, S., Faggio, C. (2018). How the marine biotoxins affect human health. *Nat. Prod. Res., 32*(6), 621-631.
[http://dx.doi.org/10.1080/14786419.2017.1329734] [PMID: 28532167]

Sobel, J., Painter, J. (2005). Illnesses caused by marine toxins. *Clin. Infect. Dis., 41*(9), 1290-1296.
[http://dx.doi.org/10.1086/496926] [PMID: 16206104]

Sonak, S., Patil, K., Devi, P., D'Souzac, L. (2018). Causes, human health impacts and control of harmful algal blooms: A comprehensive review. *Environ. Poll.Protec., 3*(1), 40-55.
[http://dx.doi.org/10.22606/epp.2018.31004]

Sousa, R. (2018). Marine snails of the genus phorcus: biology and ecology of sentinel species for human impacts on the rocky shores. In: Ray, S., (Ed.), *Biological Resources of Water.* intechopen.
[http://dx.doi.org/10.5772/intechopen.71614]

Stachowicz, J.J., Terwin, J.R., Whitlatch, R.B., Osman, R.W. (2002). Linking climate change and biological invasions: Ocean warming facilitates nonindigenous species invasions. *Proc. Natl. Acad. Sci., 99*(24), 15497-15500.
[http://dx.doi.org/10.1073/pnas.242437499]

Stauffer, B.A., Sukhatme, G.S., Caron, D.A. (2020). Physical and biogeochemical factors driving spatially heterogeneous phytoplankton blooms in nearshore waters of santa monica bay, USA. *Estuaries. Coasts., 43*(4), 909-926.
[http://dx.doi.org/10.1007/s12237-020-00704-5]

Stirling, D.J. (2001). Survey of historical new zealand shellfish samples for accumulation of gymnodimine. *N. Z. J. Mar. Freshw. Res., 35*(4), 851-857.
[http://dx.doi.org/10.1080/00288330.2001.9517047]

Stivala, C.E., Benoit, E., Aráoz, R., Servent, D., Novikov, A., Molgó, J., Zakarian, A. (2015). Synthesis and biology of cyclic imine toxins, an emerging class of potent, globally distributed marine toxins. *Nat. Prod.*

Rep., 32(3), 411-435.
[http://dx.doi.org/10.1039/C4NP00089G] [PMID: 25338021]

Tahri Joutei, L. (1998). *Gymnodinium catenatum* graham blooms on moroccan waters. In: Reguera, B., Blanco, J., Fernandez, M., Wyatt, T., (Eds.), *Proceedings of the VIII International Conference on Harmful Algae,* June 1999Vigo, Spain66-67.

Taleb, H., Vale, P., Jaime, E., Blaghen, M. (2001). Study of paralytic shellfish poisoning toxin profile in shellfish from the mediterranean shore of morocco. *Toxicon, 39*(12), 1855-1861.
[http://dx.doi.org/10.1016/S0041-0101(01)00167-2] [PMID: 11600148]

Templado, J., Rolán, E.A. (2012). new species of Phorcus (Vetigastropoda, Trochidae) from the Cape Verde Islands. *Iberus, 30*(2), 89-96.

Templado, J. (2014). Future trends of mediterranean biodiversity. In: Goffredo, S., Dubinsky, Z., (Eds.), *The Mediterranean Sea.* Springer, pp. 479-498.
[http://dx.doi.org/10.1007/978-94-007-6704-1_28]

Thessen, A.E., Patterson, D.J., Murray, S.A. (2012). The taxonomic significance of species that have only been observed once: The genus Gymnodinium (Dinoflagellata) as an example. *PLoS One, 7*(8), e44015.
[http://dx.doi.org/10.1371/journal.pone.0044015] [PMID: 22952856]

Tirado, M.C., Clarke, R., Jaykus, L.A., McQuatters-Gollop, A., Frank, J.M. (2010). Climate change and food safety: A review. *Food Res. Int., 43*(7), 1745-1765.
[http://dx.doi.org/10.1016/j.foodres.2010.07.003]

Torigoe, K., Murata, M., Yasumoto, T., Iwashita, T. (1988). Prorocentrolide, a toxic nitrogenous macrocycle from a marine dinoflagellate, Prorocentrum lima. *J. Am. Chem. Soc., 110*(23), 7876-7877.
[http://dx.doi.org/10.1021/ja00231a048]

Totti, C., Accoroni, S., Cerino, F., Cucchiari, E., Romagnoli, T. (2010). Ostreopsis ovata bloom along the conero riviera (northern adriatic Sea): Relationships with environmental conditions and substrata. *Harmful Algae, 9*(2), 233-239.
[http://dx.doi.org/10.1016/j.hal.2009.10.006]

Touzet, N., Franco, J.M., Raine, R. (2008). Morphogenetic diversity and biotoxin composition of Alexandrium (Dinophyceae) in Irish coastal waters. *Harmful Algae, 7*(6), 782-797.
[http://dx.doi.org/10.1016/j.hal.2008.04.001]

Turki, S., Dhib, A., Fertouna-Bellakhal, M., Frossard, V., Balti, N., Kharrat, R., Aleya, L. (2014). Harmful algal blooms (HABs) associated with phycotoxins in shellfish: What can be learned from five years of monitoring in Bizerte Lagoon (Southern Mediterranean Sea)? *Ecol. Eng., 67*, 39-47.
[http://dx.doi.org/10.1016/j.ecoleng.2014.03.028]

Ujević, I., Ninčević-Gladan, Ž., Roje, R., Skejić, S., Arapov, J., Marasović, I. (2010). Domoic acid--a new toxin in the Croatian Adriatic shellfish toxin profile. *Molecules, 15*(10), 6835-6849.
[http://dx.doi.org/10.3390/molecules15106835] [PMID: 20938398]

Ujević, I., Roje-Busatto, R., Ezgeta-Balić, D. (2019). Comparison of amnesic, paralytic and lipophilic toxins profiles in cockle (*Acanthocardia tuberculata*) and smooth clam (*Callista chione*) from the central Adriatic Sea (Croatia). *Toxicon, 159*, 32-37.
[http://dx.doi.org/10.1016/j.toxicon.2018.12.008] [PMID: 30659862]

Utermöhl, H. (1985). Zur vervollkomnung des quantitativen phytoplankton methodik. *Mitteilungen. Int. Verein. Limnology., 9*, 1-38.

Valbi, E., Ricci, F., Capellacci, S., Casabianca, S., Scardi, M., Penna, A. (2019). A model predicting the PSP toxic dinoflagellate *Alexandrium minutum* occurrence in the coastal waters of the NW Adriatic Sea. *Sci. Rep., 9*(1), 4166.
[http://dx.doi.org/10.1038/s41598-019-40664-w] [PMID: 30862824]

Vale, P., Sampayo, M.A.M. (2001). Domoic acid in Portuguese shellfish and fish. *Toxicon, 39*(6), 893-904.

[http://dx.doi.org/10.1016/S0041-0101(00)00229-4] [PMID: 11137551]

Vale, P., Sampayo, M.A.D.M. (2002). First confirmation of human diarrhoeic poisonings by okadaic acid esters after ingestion of razor clams (Solen marginatus) and green crabs (Carcinus maenas) in Aveiro lagoon, Portugal and detection of okadaic acid esters in phytoplankton. *Toxicon, 40*, 989e996.

Van Egmond, H.P., Aune, T., Lassus, P., Speijers, G., Waldock, M. (1993). Paralytic and diarrhoeic shellfish poisons: Occurrence in Europe, toxicity, analysis and regulation. *J. Nat. Toxins, 2*, 41-83.

Vilariño, N., Louzao, M., Abal, P., Cagide, E., Carrera, C., Vieytes, M., Botana, L. (2018). Human poisoning from marine toxins: Unknowns for optimal consumer protection. *Toxins, 10*(8), 324.
[http://dx.doi.org/10.3390/toxins10080324] [PMID: 30096904]

Ben-Gigirey, B., Botana, L.M., González, A.V., Rodríguez-Velasco, M.L. (2006). First evidence of spirolides in Spanish shellfish. *Toxicon, 48*(8), 1068-1074.
[http://dx.doi.org/10.1016/j.toxicon.2006.09.001] [PMID: 17046040]

Visciano, P., Schirone, M., Berti, M., Milandri, A., Tofalo, R., Suzzi, G. (2016). Marine Biotoxins: Occurrence, toxicity, regulatory limits and reference methods. *Front. Microbiol., 7*, 1051.
[http://dx.doi.org/10.3389/fmicb.2016.01051] [PMID: 27458445]

Vlamis A. and Katikou P. (2015). Human impact in Mediterranean coastal ecosystems and climate change: emerging toxins. In Climate Change and Marine and Freshwater Toxins. Luis M. Botana, Carmen Louzao, Natalia Vilariño (Eds.), 237-269. .

Wang, W., Ke, C. (2002). Dominance of dietary intake of cadmium and zinc by two marine predatory gastropods. *Aquat. Toxicol., 56*(3), 153-165.
[http://dx.doi.org/10.1016/S0166-445X(01)00205-3] [PMID: 11792432]

Wu, H., Yao, J., Guo, M., Tan, Z., Zhou, D., Zhai, Y. (2015). Distribution of marine lipophilic toxins in shellfish products collected from the chinese market. *Mar. Drugs, 13*(7), 4281-4295. Available from: https://www.mdpi.com/1660-3397/13/7/4281
[http://dx.doi.org/10.3390/md13074281] [PMID: 26184236]

Zhang, Y., Calabrese, E.J., Zhang, J., Gao, D., Qin, M., Lin, Z. (2020). A trigger mechanism of herbicides to phytoplankton blooms: From the standpoint of hormesis involving cytochrome b_{559}, reactive oxygen species and nitric oxide. *Water Res., 173*, 115584.
[http://dx.doi.org/10.1016/j.watres.2020.115584] [PMID: 32062224]

Zheng, Q., Klemas, V.V. (2018). Coastal ocean environment in reference module in earth systems and environmental sciences, comprehensive. *Remote Sens., 8*, 89-120.

Zhongming, Z., Linong, L., Wangqiang, Z., Wei, L. (2021). US Socio-Economic Effects of Harmful Algal Blooms. Available from: http://119.78.100.173/C666/handle/2XK7JSWQ/321591

Short-Term Bioassay Tests for Toxicity Effluents Estimation and Bio-Monitoring Uses in Aquatic Ecosystems

Sahar Karray[1] and **Monia El Bour**[1,*]

[1] *INSTM, Rue du 2 mars 1934, 2025, Salammbo, Tunisia*

Abstract: Bioassays are defined as the measurement of toxic responses upon exposure to chemicals under controlled conditions in the laboratory using cultured organisms and represent powerful tools for the assessment of environmental quality. These biological tests were pointed out as the most common methods used to assess the environmental risk, mainly in marine ecosystems besides biomarkers and biosensors. The list of different toxicity bioassays is still increasing and a large battery of different aquatic organisms is available for the measurement of organic and inorganic chemical toxicity.

The species selected for the available battery of bioassays differ taxonomically and play different roles in aquatic ecosystems. Else, they have different routes of exposure to bio-potential.

Although many bioassays provide information about the overall toxicity induced, new test systems are developed and used for the determination of specific toxicity in a number of biological pathways disrupted by contaminants.

Hereby, the list of bioassay tests used in aquatic ecosystems assessment is updated besides specific toxicity pathways for almost invertebrate and vertebrate aquatic species used to determine organic and inorganic pollutants effects.

Keywords: Aquatic ecosystems, Accute toxicity, Bioassays, Biomonitoring, Bioindicator organisms, Biochemical reponses, Chronic toxicity, Effluents, *In vivo*, Inorganic pollution, Invertebrates, Marine Pollution, Microbiological test, Organic pollution, Physiological reponses, Toxicity, Tests, Toxicity, Vertebrates.

---

* **Corresponding author Monia El Bour:** INSTM, Rue du 2 mars 1934, 2025, Salammbo, Tunisia
E-mail: monia.elbour@instm.rnrt.tn

Tamer El-Sayed Ali (Ed.)

INTRODUCTION

Recent data highlighted the consequences caused by the number of pollutants in aquatic environment organisms, which may remain recessive for several generations and exhibit major effects on the population (Arslan 2017). Multiple damages caused by these pollutants at the level of population and ecosystem, as in organ function (reproductive stages and biological diversity) are still increasing (Minguez *et al.* 2017, Di Paolo *et al.* 2015) and signs of alterations specifically detected in different levels (Amara *et al.* 2012, Zaaboub *et al.* 2015, Minguez *et al.* 2014, Martinez-Gomez *et al.* 2010). The evaluation of the complex actions and bioavailability of contaminants and the determination of their biological effects for almost unidentified substances have been realized by means of bioassays using live organisms (AL Shackleton *et al.* 2002, Almeida *et al.* 2012, Amara *et al.* 2012). Thus, all aquatic organisms exposed to many xenobiotics during their lives both from water, aquatic food chain, or sediments, and the degree of contamination appraised by chemical analyses is still insufficient to provide or estimate the deleterious effects on the biota despite the increasing number of recent studies related to chemical contamination of the environment and their toxic effects (Meijome *et al.* 2006, Mânzatu *et al.* 2012).

The major sources of these chemical substances are industrial and agricultural activities and almost all of these sources have ultimate contact with aquatic ecosystems, causing organic alteration and stress (Schintu *et al.* 2015). Since the last decencies, chemical analyses have been complemented with biological criteria from bioassays to design a more comprehensive approach to aquatic pollution assessment (Minguez *et al.*, 2014a, b). The bio-tests were previously recognized as tools of indication for potential toxicity of a certain effluent to the biological communities inhabiting the receiving waters and used to evaluate the status of natural environments against the impact of anthropogenic activities (MacDonald *et al.* 1997, Wells *et al.* 1998).

Thus, acute toxicity tests became routinely used to evaluate the quality of waters in aquatic coastal areas and to assess the toxicity of pollutants, mainly heavy metals, polyaromatic hydrocarbons, pharmaceutical compounds, or pesticides (Meinguez *et al.* 2017, Bao *et al.* 2013, Mottier *et al.* 2013, Mai *et al.* 2012).

The major guidelines provided for the use of biological effects techniques mainly for oil spill pollution monitoring were those of European, Mediterranean, or American regions (UNEP/MAP/MED POL 2004, Martinez-Gomez *et al.* 2010) and applied biological-effects techniques (bioassays and biomarkers) were used as tools to assist in determining the damage to health in marine ecosystems.

In addition, it has been largely reported that bioassays provide an integrative response for the combined effect of all pollutants present in effluents or waste discharges, with potential synergistic or antagonistic interactions. Therefore, the integrative assessment of aquatic pollution nowadays includes both chemical and biological monitoring (Fernández Méijome *et al.*, 2006, Gunjan *et al.*, 2015, Di Paolo *et al.* 2015).

Due to the importance of bioassays in aquatic monitoring, here we provide the bioassays' definition and present the most reported bioassays, mainly regarding the vertebrate and invertebrate species used.

DEFINITION OF BIOASSAY

Several definitions have been proposed for the bioassay as it responds only to the biologically available fraction and identifies toxicity in relevant marine organisms (Fernández Méijome *et al.* 2006). The most suitable definition is provided by the Water Framework Directive (Environment Agency 2002a), as a bioassay is the measurement of the toxic response upon exposure to a chemical under controlled conditions in the laboratory using cultured organisms. Thus, the ideal bioassay remains simple to emulate, regularly inter-calibrated, sensitive to a wide range of pollutants, able to utilize organisms from reliable stock, practicable, relevant, readily understood, and able to yield statistically robust data (Environment Agency 2002b). It is important to specify that a bioassay can provide predictions of environmental impacts, whereas ecological community measures only impacts after they occur, and thus, the use of bioassays for investigative monitoring is probably their most important role related to the requirements of the Directive of Environment Agency (2002a) and those recently published (Hunting *et al.* 2017).

DIFFERENT TYPES OF BIOASSAYS

Different bioassays using different species representative of the ecosystem have been used to obtain relevant information on potential ecological risks in polluted environments. Species selected for the different proposals differ by taxonomy, their roles in ecosystems, route of exposure, availability, and easy culture in the laboratory (Zovkol, 2015). Bioanalytical bioassays as biomarkers and biosensors represent suitable detection systems signaling potential damage in the environment and a wide variety of biological bioassays have been developed to evaluate different endpoints, such as sub-lethal effects, and biochemical responses (Stekenburg *et al.* 2007). In recent decades, toxicity biotesting has grown steadily as a useful tool in environmental risk assessment, including the most used direct toxicity assessment tests, water quality index, and *in vivo* and *in vivo* bio-tests (Di Paolo *et al*. 2015, Schintu, 2015).

Toxicity is defined as the adverse or harmful effect of a chemical substance upon the biological system of an organism over the suitable period (Stekenburg *et al*. 2007). Depending on the time and life duration of the specimen assessed, two types of toxicity were found: acute toxicity (short-term toxicity) and chronic toxicity (long-term toxicity). Acute toxicity is usually indicated by death, which is used in laboratory toxicity tests to assess the lethal concentration of a sample or compound (Ross 2012, Di Paolo *et al*. 2015). Previously, the EPA guidelines specified that toxicological bioassay measurements occur over a range of time scales, and short-term acute lethality tests are carried out by exposing animals to chemicals for a known time of exposure (EPA 2002). The time scale can differ depending on the objectives of the test but should be between 24 and 96 hours. Furthermore, specifically for aquatic organisms, the elements tested were introduced by mixing in a water medium, and two measures were determined based on mortality: the first is the Lethal Concentration 50 (LC50: concentration of the chemical in the water that causes death for 50% of the animals) (Eaton and Klaasen 2003). The other measurement is the No-Observed-Adverse-Effect-Concentration (NOAEC: the highest concentration at which survival is not significantly different from control). For acute toxicity tests to be statistically rigorous, a minimum of 20 test organisms should be exposed to each concentration of interest (EPA 2002, Ross 2012).

Bioassays usually measure more sublethal effects in organisms under defined conditions and can be carried out *in situ* in water bodies, or in laboratory using samples of water collected from the field (Persson 2012).

Sub-lethal methods compared to acute tests are generally more sensitive to range dilutions of contaminants expected in water bodies after discharge or dispersion, and many sub-lethal responses can be detected by physiological measurements, such as growth rates, oxygen production, while others can be manifested by behavioral, biochemical or mutagenic changes (Connon *et al*. 2012, Gunjan *et al*. 2015).

Thus, for active bio-monitoring, organisms can be exposed *in situ* and the response is measured at regular time intervals, or they can be placed at selected sites over a study area. For all sites, the results were evaluated for the same time interval (Gunjan *et al*. 2015). Commonly used bioassays are *in vitro* cell-based bioassays and *in vivo* cell-based bioassays.

In Vitro Cell-based Bioassay

In vitro bioassays offer a rapid and sensitive detection of chemical activities or combined effects of chemical substances with similar modes of action at the cellular level (Connon *et al.* 2012). These bioassays are qualified as highly sensitive and respond to many substances while having a relatively short exposure time. For these bio-tests, specific effects of chemicals were monitored on cells and can be used for high throughput screening (HTS). They measure effects at the subcellular level, such as receptor activation and DNA damage, rather than investigating cells or tissues of organisms, and through the integration of the effect of all substances with the same mode of action, *in vivo* bioassays quantify and distinguish antagonistic effects (Wernersson *et al.* 2015).

The *in vitro* biotests were developed for ethical, scientific, and economical reasons. They are a valuable tool for assessing toxicity and cell viability that specifically measures the adaptive stress response. An advantage of this approach is that *in vivo* bioassays can often be performed on many different matrices (such as concentrated extracts of surface water, sediment or pore water samples, biological tissues, passive samplers, and effluents). Additional advantages are that only small amounts of samples are generally needed and the exposure time is short compared to the time needed for an *in vivo* assay to detect a response.

In vitro bioassays comprise bacterial tests of acute toxicity (*e.g.*, Microtox, *Vibrio fischeri* assay), genotoxicity (*e.g.*, Umu test, Ames test), and reporter gene assays for measuring dioxin receptor activation (DR-CALUX) (Gomez *et al.*, 2010). The application of 3-4 *in vivo* bioassays covering different modes of action (AhR-mediated toxicity, acute toxicity, and genotoxicity) of the toxicants is recommended, particularly including the DR-CALUX. For a more detailed toxicity assessment, reporter gene assays for estrogenicity screening can be included (YES, ER-CALUX).

Microbial toxicity tests have been commonly used for the risk assessment of polluted sediments and showed high reproducibility and least variability among laboratories (Stronkhorst *et al.* 2004). Bioluminescent analysis of microbial bioassays is one of the most promising expression methods for biologically monitoring the environment as it continues to be highly sensitive to micro quantities of pollutants (Medvedeva 2009). The use of *in vivo* assays is increasing for ethical reasons to comply with the regulations on animal experimentation. The advantage of such an approach is that it can often be performed on many different matrices (concentrated extracts of surface water, sediment or pore water samples, biological tissues, passive samplers, and effluents), using small amounts of samples with short time of exposure compared to *in vivo* bioassay to detect a response (Bondarenko *et al.* 2016). Else, the effects of all substances with the same mode of action (*e.g.*, estrogen receptor binding) are detected and as integrative detection tools, *in vivo,* assays can quantify and distinguish agonistic and antagonistic effects. Such effects of several substances with the same mode of action were demonstrated using *in vivo* assays on primary cultures of *Haliotis tuberculata* hemocytes by Minguez (2017).

However, *in vitro* bioassays are critical and considered highly simplified and do not cover the complexity of organisms. Else, their predictive power for effects (at higher organizational levels) is still limited (Van Beelen *et al.*, 2003 Wernersson *et al.* 2015). Therefore, several literature studies were considered for monitoring surveys or toxicity detection, chemical analyses combined with *in vivo* and *in vivo* bioassays (Houtman *et al.*, 2004, Boundarenko *et al.* 2016).

In Vivo Cell-based Bioassay

In vivo, bio tests are commonly used to determine the adverse effects of chemicals on tissues or whole organs and organisms based on a wide variety of endpoints, including enzyme activity, cell differentiation, and exposure requirement of the whole organism to determine toxicity. These types of assays are indicative of specific endpoints relevant to human or environmental health and measure the effect of pollutants on growth, reproduction, feeding activity, and mortality (Gunjan *et al.* 2015).

Increasingly, toxicity assessment studies use standardized aquatic organisms both from freshwater and marine water ecosystems (Minagh *et al.*, 2009 Minguez *et al.*, 2014b).

Several aquatic species are still considered for toxicity monitoring assessment *via in vivo* bioassays. For almost all studies, the evaluation of a suitable candidate organism for toxicity testing purposes requires the comparison of biologically measurable responses or end-points to a range of concentrations for a common reference toxicant. With this aim, the sensitivity of various organisms can be ranked, and the most suitable species are already identified (USEPA, 1995). For practical use and selection of a specific bioassay, the candidate is designed to be sensitive for at least one of its life history stages, to potential contaminants and the field available (relatively easy to collect or abundant) and amenable to routine maintenance, culture, and rearing in the laboratory (Rand, 1995). Thus, for almost all use of developmental stages, it has been recommended that spawning should be readily induced and gametes were available available from the natural habitat (De Castro-Català *et al.* 2016).

Therefore, several *in vivo* toxicity bioassays have been developed to monitor the presence of pollutants in the aquatic environment, employing different kinds of animals including mainly vertebrate fishes and invertebrates: bivalves, sea urchins, sand dollars, and arthropods as test organisms (King and Riddle 2001; Carr and Nipper. 2003 and Shi Ling Chan. 2013). Here, we summarize the main *in vivo* bioassays according to the kind of aquatic organisms used.

Fishes In Vivo Bioassay

Fishes used for the *in vivo* bioassays are from different species, should be in good health, free from any apparent malformation, and must be held in the laboratory for at least 12 days before the test (Wenersson *et al.* 2015).

The main fish bioassay is still "the fish embryo acute toxicity (FET) test, based on individual exposure of eggs to evaluate the embryo toxicity of samples to detect contaminants (relevant for the Water Framework Directive (WFD) such as industrial chemicals, pesticides, pharmaceuticals and biocides (Wernersson *et al.* 2015). For another aim, the acute toxicity of heavy metals like cadmium and mercury was determined by static bioassay exposing the fish as reported for the Asian sea bass (*Lates calcarifer)* which is considered as important candidate fish species for brackish water aquaculture in India (Raj *et al.* 2013). To standardize a fish embryo test, the acute toxicity of metals and polycyclic aromatic hydrocarbons (PAHs) to early life stages (ELS) of turbot (*Psetta maxima*) was studied by Mhedhbi *et al.* (2010). The effect of drilling fluid (both base oil and drilling mud) was tested on three fish species (estuarine fish: *Tilapia mossambica,* marine fish: *Mugil persia,* and benthic sediment fish: *Boleophthalmus boddarti*) (Sil *et al.* 2012). Over the last 10 years, the great majority of studies performed bioassays with embryos and early larvae which allowed small-scale and low-volume experimental setups, minimized sample use and reduced workload, and used Zebrafish (*Danio rerio* tropical freshwater fish from the family of *Cyprinidae*) as vertebrate organism model broadly applied in biological sciences and is one of the most important organisms used in different research areas in genetics, developmental biology, and ecotoxicology (Di Paolo *et al.* 2015).

Invertebrate In Vivo Bioassay

Several *in vivo* bioassays are practiced on invertebrate animals. They must meet several criteria such as simplicity, speed, reproducibility, and economic cost (Ramade, 2008). These bioassays can be studied at different levels of biological organization: intracellular (enzymatic, genetic, *etc.*), cellular, population, or even at communities and then at ecosystem levels. Different endpoints can be measured: mortality, growth, development, reproduction, enzymatic activity, genotoxicity, mutagenicity, *etc.* (Calow, 1993; Forbes and Forbes, 1997). We cite here the most used tests in invertebrates:

- Aquatic ecotoxicology biotest for embryo larval bivalve development according to the T90 XP-283 norms (AFNOR, 2009).
- The daphnia test, performed on *daphnia magna*, is standardized by AFNOR under the reference NF T 90301.
- The stress-to-stress test gives an overall assessment of the state of health of organisms without indicating the type of contaminants present in the environment (Viarengo *et al.* 1995). This test assesses the ability of organisms to support anoxia. It consists of adding anoxic stress to animals already disturbed in their original environment. (Viarengo *et al.* 1995; Ladhar-Chaabouni *et al.* 2008).
- Micro-algea culture bioassays exposed to a toxic pollutant. The most widely used is the test for inhibiting the growth of the unicellular freshwater Chlorcoccale *S.capricornutum.*
- Bioassays of effects on the reproduction of terrestrial gastropod *Helix aspersa* (Gomot and Gomot, 2004).
- Growth inhibition test in terrestrial and aquatic invertebrates
- Micronucleus test for amphibians (Ramade, 2008)

The use of a single bioassay cannot take into consideration the diversity of contaminants and the multiplicity of their effects on organisms. However, using a coherent set of bioassays studied at different levels of biological organization makes it possible to establish a complete diagnosis of disturbances in vertebrates and invertebrates organisms (Naija *et al.*, 2015).

CONCLUSION

This study represents a compilation of the most important data on bioassays as powerful tools for assessing environmental quality. Thereby, the various types of bioassays, methodologies (*in vivo*/*in vitro*) as well as the species (vertebrates/ invertebrates) have been discussed in this paper. Else, we highlighted here the importance of using a coherent set of bioassays studied at different levels of biological organization for establishing a complete diagnosis of disturbances in vertebrates and invertebrates organisms.

REFERENCES

Bartram, J, Ballance, R (1996). *Water Quality Monitoring - A Practical Guide to the Design and Implementation of FreshwaterQuality Studies and Monitoring ProgrammesPublished on behalf of United Nations Environment Programme and the World Health Organization UNEP/WHO* WHO ORG.

Bondarenko, O.M., Heinlaan, M., Sihtmäe, M., Ivask, A., Kurvet, I., Joonas, E., Jemec, A., Mannerström, M., Heinonen, T., Rekulapelly, R., Singh, S., Zou, J., Pyykkö, I., Drobne, D., Kahru, A. (2016). Multilaboratory evaluation of 15 bioassays for (eco)toxicity screening and hazard ranking of engineered nanomaterials: FP7 project NANOVALID. *Nanotoxicology, 10*(9), 1229-1242.
[http://dx.doi.org/10.1080/17435390.2016.1196251] [PMID: 27259032]

Calow, P. (1993). General principales and overreview. Handbook of ecotoxicology / edited by Peter Calow.*Library Catalog* Oxford, UK ; Malden, MA : Blackwell Science.

Cherchi, A. (2012). The response of benthic foraminiferal assemblages to copper exposure: A pilot mesocosm investigation. *J. Environ. Protec., 3*, 342-352. Available from: http://www.SciRP.org/journal/jep
[http://dx.doi.org/10.4236/jep.2012.34044]

Corbi, J.J, Gorni, G.R, Correa, R.C. (2015). An evaluation of Allonais inaequalis Stephenson, 1911 (Oligochaeta: Naididae) as a toxicity test organism. *Ecotoxicol. Environ. Contam., 10*, 7-11.
[http://dx.doi.org/10.5132/eec.2015.01.02]

De Castro-Català, N., Kuzmanovic, M., Roig, N., Sierra, J., Ginebreda, A., Barceló, D., Pérez, S., Petrovic, M., Picó, Y., Schuhmacher, M., Muñoz, I. (2016). Ecotoxicity of sediments in rivers: Invertebrate community, toxicity bioassays and the toxic unit approach as complementary assessment tools. *Sci. Total Environ., 540*, 297-306.
[http://dx.doi.org/10.1016/j.scitotenv.2015.06.071] [PMID: 26118861]

Environment Agency. (2002). Role, Application and Guidance for Use of Bioassays inthe Monitoring and Management of the Water Environment – Literature Review. R&D Technical Report E1- 083/TR2.

Environment Agency. (2002). General Framework for the Use of Bioassays in the Monitoring and Management of the Water Environment. R&D Technical Report E1- 083/TR1.

Frontalini, A., Coccioni, R., Adamo, R. (2010). Computerized sperm motility analysis in toxicity bioassays: A new approach to pore water quality assessment. *Ecotoxicol. Environ. Saf., 73*, 1588-1595.
[http://dx.doi.org/10.1016/j.ecoenv.2010.05.003] [PMID: 20537390]

Forbes, V.E., Forbes, T.L. (1997). *Ecotoxicologie. Théorie et applications..* Paris: INRA Editions.

Gunjan, D., Garg, B. (2015). Toxicity tests to check water quality international research. *J.Environ. Sci., 4*(11), 87-90.

Houtman, C.J., Cenijn, P.H., Marja, T.H., Harmers, T., Lamoree, J.L., Albertinka, J.M. (2004). Toxicological profiling of sediments using *in vitro* bioassays, with emphasis on endocrine disruption. *Environ. Toxicol. Chem., 23*(1), 32-40.
[http://dx.doi.org/10.1897/02-544]

Hunting, E.R., de Jong, S., Vijver, M.G. (2017). Assessment of monitoring tools and strategies safeguarding aquatic ecosystems within the European water framework Leiden University, Institute of Environmental Sciences. *Conserv. Biol.*

Mânzatu, C., Nagy, B., Ceccarini, A., Iannelli, R., Giannarelli, S., Majdik, C. (2015). Laboratory tests for the phytoextraction of heavy metals from polluted harbor sediments using aquatic plants. *Mar. Pollut. Bull., 101*(2), 605-611.
[http://dx.doi.org/10.1016/j.marpolbul.2015.10.045] [PMID: 26515993]

Méijome, I.F., Fernández, S., Beiras, R. (2006). Assessing the toxicity of sandysediments six months after the Prestige oil spill by means of the sea-urchin embryo larval bioassay. *Thalassas, 22*(2), 45-50.

Minguez, L., Di Poi, C., Farcy, E., Ballandonne, C.L., Benchouala, A., Bojic, C.L., Leguille, C., Costil, K., Serpentini, A, Lebel, J.M, Halm-Lemeille, MP (2014). Comparison of the sensitivity of seven marine and freshwater bioassays as regards antidepressant toxicity assessment. *Ecotoxicology, 23*(9), 1744-1754. [http://dx.doi.org/10.1007/s10646-014-1339-y]

Naija. (2016). *Physiological and molecular evaluation of the toxicity of heavy metals on a species of Blennid from the Tunisian coasts. Doctoral thesis* University of Monastir.

Ramade, F. (2008). *Introduction to Ecotoxicology. Foundations and applications*, Lavoisier.

Raj, V.M, Thirunavukkarasu, A.R., Kailasam, M, Muralidhar, M, Subburaj, M (2013). Acute toxicity bioassays of cadmium and mercury on the juveniles of asian seabass lates calcarifer (bloch). *Ind. J. Sci.Tech., 6*(4).

Ross. (2012). Responses to chemical exposure by foraminifera: Distinguishing dormancy from mortality Benjamin James Ross. University of South Florida.

Schintu, M., Carla, B., François, G., Alessandro, M., Barbara, M., Angelo, I. (2015). Interpretation of coastal sediment quality based on trace metal and PAH analysis, benthic foraminifera,and toxicity tests (Sardinia, Western Mediterranean). *Mar. Pollut. Bull., 94*(1-2), 72-83.

Stronkhorst, J., Ciarelli, S., Schipper, C.A., Postma, J.F., Dubbeldam, M., Vangheluwe, M., Brils, J.M., Hooftman, R. (2004). Inter-laboratory comparison of five marine bioassays for evaluating the toxicity of dredged material. *Aquat. Ecosyst. Health Manage., 7*(1), 147-159. [http://dx.doi.org/10.1080/14634980490281579]

Stronkhorst, J., Ciarelli, S., Schipper, J. F, Postma, M., Dubbeldam, M., Vangheluwe, J., Persson, L. (2013). Screening methods for aquatic toxicity of surfactants. Master of Science Thesis in the Master Degree Programme Materials and Nanotechnology. Department of Chemistry and Biotechnology. Division of Applied Surface Chemistry. Chalmers university of technology. Gothenburg, Sweden.

Steken, B. (2007). *Proc. of the 3rd IASME/WSEAS Int. Conf. on Energy, Environment, Ecosystems and Sustainable Development.*Agios Nikolaos, GreeceJuly 24-26, 2007

Van der Oost, R., Beyer, J., Vermeulen, N.P.E. (2003). Fish bioaccumulation and biomarkers in environmental risk assessment: A review. *Environ. Toxicol. Pharmacol., 13*(2), 57-149. [http://dx.doi.org/10.1016/S1382-6689(02)00126-6] [PMID: 21782649]

Van Beelen, P. (2003). A review on the application of microbial toxicity tests for deriving sediment quality guidelines. *Chemosphere, 53*(8), 795-808. [http://dx.doi.org/10.1016/S0045-6535(03)00716-1] [PMID: 14505700]

Wernersson, S., Carere, M., Magg, C., Kase, R. (2015). The European technical report on aquatic effect-based monitoring tools under the water framework directive. *Environ. Sci. Eur., 27*(7). [http://dx.doi.org/10.1186/s12302-015-0039-4]

CHAPTER 3

The Challenge of Microplastics in Aquatic Ecosystem: A Review of Current Consensus and Future Trends of the Effect on the Fish

Tamer El-Sayed Ali[1,*]

[1] *Department of Oceanography, Faculty of Science, Alexandria University, Alexandria, Egypt*

Abstract: In recent decades, the prevalence of plastics in the marine environment has increased and is amongst the most pervasive problems affecting the marine environment globally. Numerous studies have documented microplastic ingestion by marine species with more recent investigations focusing on the secondary impacts of microplastic ingestion on ecosystem processes. However, few studies so far have examined microplastic ingestion by mesopelagic fish which are one of the most abundant pelagic groups in the oceans and their vertical migrations are known to contribute significantly to the rapid transport of carbon and nutrients to the deep sea. Therefore, any ingestion of microplastics by mesopelagic fish may adversely affect this cycling and may aid in the transport of microplastics from surface waters to the deep-sea benthos.

Microplastics are ubiquitous in the marine environment and are increasingly contaminating species in the marine ecosystem and the food chain, including food stuffs intended for human consumption. The effects of microplastics on aquatic organisms are currently the subject of intense research. Here, we provide a critical perspective on published studies of microplastic ingestion by aquatic biota. We summarize the available research on Microplastic presence, behavior, and effects on aquatic organisms monitored in the field and laboratory studies of the ecotoxicological consequences of microplastic ingestion.

Finally, researchers plan further studies to learn more about how these fish are ingesting and spreading microplastics. It will be particularly interesting to see whether the fish ingest these microplastics directly as mistaken prey items, or whether they ingest them through eating prey species, which have previously ingested the microplastics. Also, there is a need to understand the mechanism of action and ecotoxicological effects of environmentally relevant concentrations of microplastics on aquatic organism health.

* **Corresponding author Tamer El-Sayed Ali:** Department of Oceanography, Faculty of Science, Alexandria University, Alexandria, Egypt; E-mail: tameraly@yahoo.com

Keywords: Biota, Ecotoxicology, Marine ecosystem, Microplastics.

INTRODUCTION

Despite scientific concern and the increasing worldwide contamination of aquatic ecosystems with hundreds of natural and industrial chemical compounds, research on the pollution of marine environments from plastic is scarce. There is an enormous quantity of plastic litter in the aquatic environment. Approximately, 8.3 billion metric tons of plastic had been produced in 2017, and its production is being steadily increased yearly (Geyer *et al.*, 2017). Up to 10% of the plastic produced each year worldwide terminates in the aquatic environment, where it persists and accumulates (Jambeck *et al.*, 2015), thus being ubiquitous in the environment. By 2050, plastic stocks in the ocean will equal and may surpass fish stocks (Gallo *et al.*, 2018). Plastics do not last forever and the marine environment is suitable for their degradation, meaning fragmenting the original plastic pieces, turning the oceans into a soup of microplastics. With the time, the abundance of micro- and nano-plastic particles will be considerably higher than the number of plankton. Thus, it is one of the environmental problems facing humanity (Thompson *et al.*, 2009; Faggio *et al.*, 2018; Strungaru *et al.*, 2018; Savoca *et al.*, 2019a).

These particles may be accidentally and deliberately ingested by marine organisms. Therefore, they may cause significant impacts on these organisms, such as inflammation, mal-feeding, and weight loss. Microplastic contamination may also spread from one organism to another when prey are ingested by predators. As the fragments can bind to chemical pollutants, these associated toxins could accumulate in predators. Moreover, laboratory studies indicate that micro- and nano-plastics can cause various negative effects in fish such as physical damage, abnormal behavior, change in lipid metabolism, as well as cytotoxicity. Thus these pollutants are of high concern regarding the ecosystem, animal state, and human health (WHO, 2010). They are globally dispersed in marine, freshwater, and terrestrial ecosystems (Ng *et al.*, 2018). Additionally, they can be transported for long distances mainly through long-range air circulation, water currents, and contaminated biota. In general, animals and humans may be exposed to both substances through contaminated food, water, air, and soil (Horton *et al.*, 2017; Barboza *et al.*, 2018; Lehner *et al.*, 2019). Therefore, this review aims to raise awareness on the topic of the potential effects of micro-and nanoplastics ingestion by fish and investigate if microplastics can interact positively or negatively with the eco-toxicological effects of other contaminants. There may be synergistic effects of microplastics and other contaminants on organism health, or they may function as a transport vector for other contaminants. Moreover, this review aims to increase overdue attention to such a class of chemicals, a topic that has not been sufficiently investigated.

SOURCES AND DISTRIBUTION

In the marine ecosystem, microplastics are a heterogeneous group of particles that vary in shape, size, and chemical composition. They are found in sediments/sea floor, on the sea surface, in the water column, and also in wildlife. The most commonly manufactured plastics are polyethylene and polypropylene (US EPA, 2018). Microplastics could be defined as tiny fragments (less than 5mm in size) of degraded plastic, synthetic fibers, and plastic beads that have accumulated in the marine environment following decades of pollution (Blair *et al.*, 2017). While nanoplastic particles are defined as plastic particles ranging from 1 to 100 nm in size (Weis *et al.*, 2015). Microplastics are often classified into primary and secondary types. Primary microplastics were originally produced to be < 5 mm in size, while secondary microplastics result from the breakdown of larger items. Microbeads which are used in personal care products, industrial abrasives, and pre-production plastic pellets used to make larger plastic items are the main sources of primary microplastics (Anderson *et al.*, 2016). Sources of secondary microplastics include microfibers from textiles, tire dust, and larger plastic items that degrade and consequently fragment into microplastic particles, mostly due to natural forces like sunlight and wave action (Duis and Coors, 2016). The extent of plastic degradation depends on factors including polymer type, age, and environmental conditions like weathering, temperature, irradiation, and pH (Akbay and Özdemir, 2016). Additionally, marine microplastics would continue to increase as larger plastic litter degrades into secondary microplastics (Anderson *et al.*, 2016).

PHYSICAL AND CHEMICAL PROPERTIES

Plastics are cheap, light in weight, strong, durable, and corrosion-resistant materials, with high electrical and thermal insulation properties. The diversity of polymers and the versatility of their properties are used to make a variety of products that bring technological advances, energy savings, and many other benefits (Andrady and Neal, 2009; Savoca *et al.* 2019b).

Microplastics in the marine environment are typically found as pellets, fragments, or fibers and are composed of diverse polymers. Such microplastics are commonly denser than seawater and are expected to sink to the seabed. These include polyester, polyamide, and polyvinyl chloride (PVC). Others are lighter than seawater and are often found floating on the water surface. These include polyethylene, polypropylene, and polystyrene (Smith *et al.*, 2018).

As plastics are polymers, they are composed of monomers joined to make the polymer structure. During production, plastic is processed with additives to provide specific properties (Lithner, 2011). Such additives account for approximately 4% of the weight of microplastics and may include plasticizers, flame retardants, pigments, antimicrobial agents, ultraviolet stabilizers, heat stabilizers, fillers, and flame retardants such as polybrominated diphenyl ethers (PBDEs) (EFAS, 2016, El-Sayed Ali and Kheirallah, 2016).

POTENTIAL EFFECTS OF CHEMICAL ADDITIVES

The previously mentioned additives, especially the chemical ones can cause toxic and hazardous effects on human health and the environment when they leach from plastic polymers. As plastics progressively degrade, additive chemicals are expected to leach. For organisms that have directly ingested microplastics, the uptake rate of chemicals by an organism's gastrointestinal tract is primarily influenced by the chemical fugacity gradient between the organisms' tissues and the plastic, the gut retention time of the microplastics, and the material-specific kinetic factors (Mato *et al.*, 2001).

In addition to chemical additives which are associated with plastic debris, microplastics in the ocean accumulate persistent organic pollutants (POPs) such as polychlorinated biphenyls (PCBs), polycyclic aromatic hydrocarbons (PAHs), and organochlorine pesticides such as DDT. These have a greater affinity for plastic than water, being higher in microplastics than in surrounding water. PBDEs are human-made flame-retardant chemicals. PBDEs enter the marine environment mainly *via* discarded or leaked consumer goods or municipal wastes (Andrady, 2011; Rochman *et al.*, 2013; Fiorino *et al.*, 2018; Stara *et al.*, 2019; Stara *et al.*, 2020).

The global distribution of chemicals in the marine environment may affect environmental and human health, but microplastics do not represent the only exposure pathway. Microplastics may represent a relatively small contributor to the total risk as there are many other sources for chemical exposure (Fossi *et al.*, 2012). For example, the total dietary intake of PCBs from microplastics is likely minimal compared to that from other sources (Jambeck *et al.*, 2015). For other chemicals, such as bisphenol A (BPA) or PBDEs, sources of exposure may be limited to or originate from microplastic degradation.

Moreover, the ability of microplastics to accumulate POPs raises concern that microplastics could transfer hazardous POPs to marine animals and subsequently humans (Rochman *et al.*, 2013). Direct exposures to POPs and other chemicals associated with microplastics may affect biological systems and pose threats to humans and animals, even at low doses (Science for Environmental Policy, 2011).

Several studies demonstrate increased toxicity from the combination of microplastics and associated chemicals (Jartun *et al.*, 2009, Hirai *et al.*, 2011). In animals, the quantity of chemicals from microplastics is suspected to be minimal compared to that from other components of the diet (Jambeck *et al.*, 2015). Microplastics and their constituents may exert localized particle toxicity, but chronic exposure producing a cumulative effect is of higher concern.

ACCUMULATION OF PLASTIC WASTES IN THE AQUATIC ECOSYSTEMS

The widespread distribution of microplastics in aquatic ecosystems causes a wide range of aquatic organisms highly susceptible to these emerging pollutants (Lusher *et al.*, 2017). Plastic ingestion occurs across various taxa within different trophic levels, including marine mammals, invertebrates, and fish (Gall and Thompson, 2015).

There are several ways in which organisms can accumulate microplastics. Animals exposed to microplastics may incorporate them *via* their gills (Watts *et al.*, 2014) and gut (de Sá *et al.*, 2015). Ingestion is a common interaction between living organisms and microplastics. The fate and impact of microplastics and their associated chemicals vary across species and environments (Jambeck *et al.*, 2015). Ingestion may be due to an inability to differentiate microplastics from prey or ingestion of organisms of lower trophic levels containing these particles as plankton-containing microplastics (do Sul and Costa, 2014). Microplastics may also adhere directly to organisms (Cole *et al.*, 2013). When these particles are found concentrated in an organism's digestive tract, they translocate from the intestinal tract to the circulatory system or the surrounding tissue (Lusher *et al.*, 2017). With a preference for smaller particles, micro- and nano plastics can persist in the animal's body (Murray and Cowie, 2011; Cauwenberghe and Janssen, 2014).

Ecotoxicological Studies on Aquatic Organisms

Ecotoxicological studies of microplastics have been conducted predominantly using marine (77%) and freshwater organisms (23%) including fish, crustaceans, mollusks, annelids, reptiles, *etc.* The limitations of ecotoxicological knowledge on the behavior of microplastics in the freshwater environment have been discussed by several authors (Wagner *et al.*, 2014; Eerkes-Medrano *et al.*, 2015; Phuong *et al.*, 2016). This knowledge gap is of high concern since freshwater organisms are directly and highly affected by terrestrial runoff, wastewater, and other discharges highly containing significant levels of microplastics and other contaminants (Dris *et al.*, 2015).

Fish are the most commonly studied group (25%) in the marine environment, followed by mollusks (15%), crustacea (19%), annelids (6%), mammals and echinoderms (both 3%), birds and cnidaria (both 2%), Porifera, reptiles and rotifers (all < 1%). Multiple freshwater studies only exist for fish and small crustaceans, and very limited studies are on freshwater birds, amphibians, annelids, and rotifers.

On the experimental level, the most studied species is the small freshwater planktonic crustacean *Daphnia magna* (12%) (Booth *et al.*, 2016; Jemec *et al.*, 2016; Ma *et al.*, 2016; Ogonowski *et al.*, 2016; Rehse *et al.*, 2016; Kim *et al.*, 2017; Mattson *et al.*, 2017). The number of studies of *D. magna* may reflect its widespread usage in ecotoxicological studies. Moreover, closely related species may show differences in response to microplastic exposure. Jaikumar *et al.* (2018) showed that three *Daphnia* species showed different responses to primary and secondary microplastic exposure and different interactive effects of microplastic exposure and thermal stress. While the authors noted that their study should be interpreted with caution due to the high concentrations of microplastics used.

Another commonly studied species in ecotoxicological studies is the freshwater zebrafish (*Danio rerio*) (7%) (El-Sayed Ali and Legler, 2011; Lu *et al.*, 2016; Chen *et al.*, 2017; Sleight *et al.*, 2017; Veneman *et al.*, 2017; Lei *et al.*, 2018), the common goby, a marine fish *Pomatoschistus microps* (6%) (Oliveira *et al.*, 2013; de Sá *et al.*, 2015; Ferreira *et al.*, 2016; Fonte *et al.*, 2016), a mollusc (*Mytilus edulis*; 5%) (De Witte *et al.*, 2014, Van Cauwenberghe and Janssen, 2014, Van Cauwenberghe *et al.*, 2015), and the annelid lugworm *Arenicola marina* (5%) (Besseling *et al.*, 2013; Van Cauwenberghe *et al.*, 2015; Green *et al.*, 2016), it is not obvious to what extent results based on studies of marine organisms can be applied to freshwater species and *vice versa*.

Respecting the marine systems, microplastics are consumed by more than two hundred species, however, the harmful effects of microplastic ingestion in marine organisms, including corals, are not well understood yet. The effects of microplastics on fish species are still under debate. Few species have shown toxicological or physical impact due to the ingestion of microplastics, such as liver toxicity or alterations of intestinal tissues. The European seabass (*Dicentrarchus labrax*) is potentially exposed to the ingestion of microplastics both in its natural habitat and in the aquaculture facilities. *Via* Real-Time PCR, a variation of four different early warning signals in the liver of the European sea bass (*Dicentrarchus labrax*) was exposed for ninety days to virgin and marine polluted PVC microplastics supplemented with food. This study represents one of the first studies on the effects of exposure to virgin and marine polluted PVC microplastics on an edible species, which shows an early warning signal on the chemical and physical hepatic stress on that species. Further data are needed to better understand the role of the partitioning of chemicals from and to microplastics and the related effects on fish and consequently on human health.

In a recent study, Wieczorek *et al.* (2018) found that 73% of mesopelagic fish caught in the Northwest Atlantic had microplastics in their stomachs. Typically living at depths of 200-1000 meters, these fish could spread microplastic pollution throughout the marine ecosystem, by carrying microplastics from the surface down to deeper waters. They are also prey for fish eaten by humans, meaning that microplastics could indirectly contaminate our food supply.

Three commercially demersal fish species - the lesser spotted dogfish *Scyliorhinus canicula*, the European hake *Merluccius merluccius* and the red mullet *Mullus barbatus* were collected from the Spanish Atlantic and Mediterranean coasts. A number of 212 fish were examined, 72 dogfish, 12 hakes, and 128 red mullets for microplastic ingestion. The percentage of fish with microplastics was 17.5% (15.3% dogfish, 18.8% red mullets, and 16.7% hakes), averaging 1.56 items per fish, and the size of the microplastics ranged from 0.38 to 3.1 mm. These species are used currently for monitoring marine pollution and may be as suitable candidates for monitoring spatial and temporal trends of ingested litter (Bellas *et al.*, 2016).

Scomber japonicus registered the highest mean of ingested microplastics, suggesting its potential role as indicator species to monitor and investigate trends in ingested litter off the Portuguese coast.

In vivo studies have demonstrated that nanoplastics can translocate to all organs. Evidence is evolving regarding relationships between micro- and nano-plastic exposure, toxicology, and human health.

Interactive Ecotoxicological Effects of Microplastics with other Contaminants

There are a vast number of studies on the combined effects of microplastics with other environmental pollutants. These studies are motivated by the situation found in the environment where organisms are simultaneously exposed to various contaminants at the same time. Due to their lipophilic nature, microplastics have the potential to adsorb persistent organic pollutants present in contaminated regions, which may increase their harmful effect once assimilated by organisms. In a study realized by O'Donovan *et al.* (2018), environmentally relevant concentrations of contaminants (BaP-16.87 ± 0.22 µg g^{-1} and PFOS-70.22 ± 12.41 µg g^{-1}) were adsorbed to microplastics to evaluate the potential role of plastic particles as a source of contamination once ingested. A multi-biomarker assessment was conducted on the gills, digestive gland, and hemolymph of clams (*Scrobicularia plana*) to clarify the effects of exposure. Results suggest a potential mechanical injury of gills caused by ingestion of microplastics that may also affect the analyzed biomarkers. The digestive gland seems less affected by mechanical damage caused by virgin microplastic exposure, with the MPs-adsorbed BaP and PFOS exerting a negative influence on the assessed biomarkers in this tissue.

Browne *et al.* (2013) observed augmented mortality of the annelid worms when exposed to PVC microplastics and the antibiotic triclosan. Fonte *et al.* (2016) showed a wide range of effects including neurotoxicity and changes in enzyme activity when *P. microps* were exposed to microplastics and the antibiotic celafaxin. Accumulation of microplastic-associated contaminants may increase the potential risk of accumulation of contaminant for higher trophic levels including in humans (Oliveira *et al.*, 2013; Luis *et al.*, 2015). The combined effects of microplastics and any environmental contaminants, after short and long-term exposure periods are still poorly known (Wright *et al.*, 2013). There are limited studies on the capability of microplastics to adsorb other contaminants inside the organisms, which may represent a positive outcome. However, there are likely to be significant analytical challenges accompanying such studies. Moreover, there is a lack of information about the interactive effects of microplastics and other pollutants on freshwater organisms.

DEGRADATION OF MARINE PLASTICS

Plastic is persistent in the marine environment because it is manufactured to be durable. Still, plastic polymers can be degraded slowly by microorganisms (biodegradation), heat (thermal degradation), oxidation (thermooxidative degradation), light (photodegradation), or hydrolysis. The rate and extent of plastic degradation are determined based on the present environmental variables.

Moreover, the size of such plastics plays a critical role in their biological fate within the marine environment. Impacts on the marine biota may vary across the size spectrum of microplastics. Large microplastics (2–5 mm) may take more time to pass from the stomach of organisms, having the potential to be retained in the digestive system. Toxicant adsorption, dependent on polymer type, may occur with increased time of exposure to plastics. Ingestion and digestion may occur with particles in the upper end of the size spectrum (1–2 mm) (Lusher *et al.*, 2017). Small marine invertebrates have been shown to ingest active particles less than 20 µm (Lee *et al.*, 2013). Smaller size microplastics have larger effects on organisms at the cellular level. In the micro- to the nanometer range, microplastics have been shown to translocate and pass into cellular membranes, including the hemolymph (Ribeiro *et al.*, 2017), and the lysosomal system (von Moos *et al.*, 2012) of marine invertebrates.

CONCLUSION AND FUTURE PERSPECTIVES

The accumulation of microplastics in the aquatic environment is an undeniable fact. Additionally, a large number of organisms are exposed to these particles. This exposure may cause a variety of effects and threats to those organisms and consequently to humans. The potentially deleterious effects of microplastics on aquatic biota have been recognized by scientists in the last years focusing basically on marine biota. However, the effects of MPs on freshwater organisms are much less known. Overall, results suggest a knowledge gap on the effects of microplastics on other organisms beyond fish. More research is required to confirm this issue and more attention should be paid to such a class of chemicals.

ACKNOWLEDGEMENTS

This work is dedicated to Prof. Altaf Ezzat, who passed away in 2018. The author is grateful to his colleagues, who helped him with data and information collection.

REFERENCES

Akbay, İ.K., Özdemir, T. (2016). Monomer migration and degradation of polycarbonate *via* UV-C irradiation within aquatic and atmospheric environments. *J. Macromol. Sci. Part A Pure Appl. Chem., 53*(6), 340-345.
[http://dx.doi.org/10.1080/10601325.2016.1165999]

Anderson, AG, Grose, J, Thompson, RC, Wyles, KJ (2016). Microplastics in personal care products: Exploring perceptions of environmentalists, beauticians and students. *Mar. Poll. Bull., 15-113*(1-2), 454-460.

Andrady, A.L., Neal, M.A. (2009). Applications and societal benefits of plastics. *Philos. Trans. R. Soc. Lond. B Biol. Sci., 364*(1526), 1977-1984.
[http://dx.doi.org/10.1098/rstb.2008.0304] [PMID: 19528050]

Andrady, A.L. (2011). Microplastics in the marine environment. *Mar. Pollut. Bull., 62*(8), 1596-1605.
[http://dx.doi.org/10.1016/j.marpolbul.2011.05.030] [PMID: 21742351]

Barboza, L.G.A., Dick Vethaak, A., Lavorante, B.R.B.O., Lundebye, A.K., Guilhermino, L. (2018). Marine microplastic debris: An emerging issue for food security, food safety and human health. *Mar. Pollut. Bull., 133*(133), 336-348.
[http://dx.doi.org/10.1016/j.marpolbul.2018.05.047] [PMID: 30041323]

Besseling, E., Wegner, A., Foekema, E.M., van den Heuvel-Greve, M.J., Koelmans, A.A. (2013). Effects of microplastic on fitness and PCB bioaccumulation by the lugworm *Arenicola marina* (L.). *Environ. Sci. Technol., 47*(1), 593-600.
[http://dx.doi.org/10.1021/es302763x] [PMID: 23181424]

Besseling, E., Wang, B., Lürling, M., Koelmans, A.A. (2014). Nanoplastic affects growth of *S. obliquus* and reproduction of *D. magna*. *Environ. Sci. Technol., 48*(20), 12336-12343.
[http://dx.doi.org/10.1021/es503001d] [PMID: 25268330]

Bellas, J, Martínez-Armental, J, Martínez-Cámara, A, Besada, V, Martínez-Gómez, C (2016). Ingestion of

microplastics by demersal fish from the spanish atlantic and mediterranean coasts. *Mar. Poll. Bull., 15-109*(1), 55-60.
[http://dx.doi.org/10.1016/j.marpolbul.2016.06.026]

Blair, R.M., Waldron, S., Phoenix, V., Gauchotte-Lindsay, C. (2017). Micro- and nanoplastic pollution of freshwater and wastewater treatment systems. Springer Science Reviews/ Analytical. *Springer Sci. Rev., 5*(1-2), 19-30.
[http://dx.doi.org/10.1007/s40362-017-0044-7]

Booth, A.M., Hansen, B.H., Frenzel, M., Johnsen, H., Altin, D. (2016). Uptake and toxicity of methylmethacrylate-based nanoplastic particles in aquatic organisms. *Environ. Toxicol. Chem., 35*(7), 1641-1649.
[http://dx.doi.org/10.1002/etc.3076] [PMID: 26011080]

Browne, M.A., Niven, S.J., Galloway, T.S., Rowland, S.J., Thompson, R.C. (2013). Microplastic moves pollutants and additives to worms, reducing functions linked to health and biodiversity. *Curr. Biol., 23*(23), 2388-2392.
[http://dx.doi.org/10.1016/j.cub.2013.10.012] [PMID: 24309271]

Chen, Q., Gundlach, M., Yang, S., Jiang, J., Velki, M., Yin, D., Hollert, H. (2017). Quantitative investigation of the mechanisms of microplastics and nanoplastics toward zebrafish larvae locomotor activity. *Sci. Total Environ., 584-585*, 1022-1031.
[http://dx.doi.org/10.1016/j.scitotenv.2017.01.156] [PMID: 28185727]

Cole, M., Lindeque, P., Fileman, E., Halsband, C., Goodhead, R., Moger, J., Galloway, T.S. (2013). Microplastic ingestion by zooplankton. *Environ. Sci. Technol., 47*(12), 6646-6655.
[http://dx.doi.org/10.1021/es400663f] [PMID: 23692270]

De Witte, B., Devriese, L., Bekaert, K., Hoffman, S., Vandermeersch, G., Cooreman, K., Robbens, J. (2014). Quality assessment of the blue mussel (*Mytilus edulis*): Comparison between commercial and wild types. *Mar. Pollut. Bull., 85*(1), 146-155.
[http://dx.doi.org/10.1016/j.marpolbul.2014.06.006] [PMID: 24969855]

Dris, R., Gasperi, J., Rocher, V., Saad, M., Renault, N., Tassin, B. (2015). Microplastic contamination in an urban area: A case study in Greater Paris. *Environ. Chem., 12*(5), 592-599.
[http://dx.doi.org/10.1071/EN14167]

Duis, K., Coors, A. (2016). Microplastics in the aquatic and terrestrial environment: Sources (with a specific focus on personal care products), fate and effects. *Environ. Sci. Eur., 28*(1), 2.
[http://dx.doi.org/10.1186/s12302-015-0069-y] [PMID: 27752437]

Eerkes-Medrano, D., Thompson, R.C., Aldridge, D.C. (2015). Microplastics in freshwater systems: A review of the emerging threats, identification of knowledge gaps and prioritisation of research needs. *Water Res., 75*, 63-82.
[http://dx.doi.org/10.1016/j.watres.2015.02.012] [PMID: 25746963]

EFAS Panel on Contaminants in the Food Chain (CONTAM). (2016). Presence of microplastics and nanoplastics in food, with particular focus on seafood. *Europ. Food. Safety. Autho. J., 14*(6).

El-Sayed Ali, T., Legler, J. (2011). Developmental toxicity of nonylphenol in zebrafish (*Danio rerio*) embryos. *Indian J. Geo-Mar. Sci., 40*(4), 509-515.

El-Sayed Ali, T., Kheirallah, N. (2016). Toxic effects of phenolic metabolite 6-OH-BDE on zebrafish (*Danio rerio*) embryos. *Indian J. Geo-Mar. Sci., 45*(5), 658-665.

Faggio, C., Tsarpali, V., Dailianis, S. (2018). Mussel digestive gland as a model tissue for assessing xenobiotics: An overview. *Sci. Total Environ., 636*, 220-229.
[http://dx.doi.org/10.1016/j.scitotenv.2018.04.264] [PMID: 29704717]

Ferreira, P., Fonte, E., Soares, M.E., Carvalho, F., Guilhermino, L. (2016). Effects of multi-stressors on juveniles of the marine fish *Pomatoschistus microps*: Gold nanoparticles, microplastics and temperature. *Aquat. Toxicol., 170*, 89-103.

[http://dx.doi.org/10.1016/j.aquatox.2015.11.011] [PMID: 26642093]

Fonte, E., Ferreira, P., Guilhermino, L. (2016). Temperature rise and microplastics interact with the toxicity of the antibiotic cefalexin to juveniles of the common goby (*Pomatoschistus microps*): Post-exposure predatory behaviour, acetylcholinesterase activity and lipid peroxidation. *Aquat. Toxicol., 180*, 173-185.
[http://dx.doi.org/10.1016/j.aquatox.2016.09.015] [PMID: 27721112]

Fiorino, E., Sehonova, P., Plhalova, L., Blahova, J., Svobodova, Z., Faggio, C. (2018). Effects of glyphosate on early life stages: Comparison between *Cyprinus carpio* and *Danio rerio*. *Environ. Sci. Pollut. Res. Int., 25*(9), 8542-8549.
[http://dx.doi.org/10.1007/s11356-017-1141-5] [PMID: 29313199]

Fossi, M.C., Panti, C., Guerranti, C., Coppola, D., Giannetti, M., Marsili, L., Minutoli, R. (2012). Are baleen whales exposed to the threat of microplastics? A case study of the Mediterranean fin whale (*Balaenoptera physalus*). *Mar. Pollut. Bull., 64*(11), 2374-2379.
[http://dx.doi.org/10.1016/j.marpolbul.2012.08.013] [PMID: 22964427]

Gall, S.C., Thompson, R.C. (2015). The impact of debris on marine life. *Mar. Pollut. Bull., 92*(1-2), 170-179.
[http://dx.doi.org/10.1016/j.marpolbul.2014.12.041] [PMID: 25680883]

Geyer, R., Jambeck, J.R., Law, K.L. (2017). Production, use, and fate of all plastics ever made. *Sci. Adv., 3*(7), e1700782.
[http://dx.doi.org/10.1126/sciadv.1700782] [PMID: 28776036]

Hirai, H., Takada, H., Ogata, Y., Yamashita, R., Mizukawa, K., Saha, M., Kwan, C., Moore, C., Gray, H., Laursen, D., Zettler, E.R., Farrington, J.W., Reddy, C.M., Peacock, E.E., Ward, M.W. (2011). Organic micropollutants in marine plastics debris from the open ocean and remote and urban beaches. *Mar. Pollut. Bull., 62*(8), 1683-1692.
[http://dx.doi.org/10.1016/j.marpolbul.2011.06.004] [PMID: 21719036]

Green, D.S., Boots, B., Sigwart, J., Jiang, S., Rocha, C. (2016). Effects of conventional and biodegradable microplastics on a marine ecosystem engineer (*Arenicola marina*) and sediment nutrient cycling. *Environ. Pollut., 208*(Pt B), 426-434.
[http://dx.doi.org/10.1016/j.envpol.2015.10.010] [PMID: 26552519]

Horton, A.A., Walton, A., Spurgeon, D.J., Lahive, E., Svendsen, C. (2017). Microplastics in freshwater and terrestrial environments: Evaluating the current understanding to identify the knowledge gaps and future research priorities. *Sci. Total Environ., 586*, 127-141.
[http://dx.doi.org/10.1016/j.scitotenv.2017.01.190] [PMID: 28169032]

Ivar do Sul, J.A., Costa, M.F. (2014). The present and future of microplastic pollution in the marine environment. *Environ. Pollut., 185*, 352-364.
[http://dx.doi.org/10.1016/j.envpol.2013.10.036] [PMID: 24275078]

Jambeck, J.R., Geyer, R., Wilcox, C., Siegler, T.R., Perryman, M., Andrady, A., Narayan, R., Law, K.L. (2015). Plastic waste inputs from land into the ocean. *Science, 347*(6223), 768-771.
[http://dx.doi.org/10.1126/science.1260352] [PMID: 25678662]

Jartun, M., Ottesen, R.T., Steinnes, E., Volden, T. (2009). Painted surfaces: Important sources of polychlorinated biphenyls (PCBs) contamination to the urban and marine environment. *Environ. Pollut., 157*(1), 295-302.
[http://dx.doi.org/10.1016/j.envpol.2008.06.036] [PMID: 18706746]

Jaikumar, G., Baas, J., Brun, N.R., Vijver, M.G., Bosker, T. (2018). Acute sensitivity of three Cladoceran species to different types of microplastics in combination with thermal stress. *Environ. Pollut., 239*, 733-740.
[http://dx.doi.org/10.1016/j.envpol.2018.04.069] [PMID: 29723823]

Jemec, A., Horvat, P., Kunej, U., Bele, M., Kržan, A. (2016). Uptake and effects of microplastic textile fibers on freshwater crustacean *Daphnia magna*. *Environ. Pollut., 219*, 201-209.
[http://dx.doi.org/10.1016/j.envpol.2016.10.037] [PMID: 27814536]

Khan, F.R., Syberg, K., Shashoua, Y., Bury, N.R. (2015). Influence of polyethylene microplastic beads on the

uptake and localization of silver in zebrafish (Danio rerio). *Environ. Pollut., 206*, 73-79.
[http://dx.doi.org/10.1016/j.envpol.2015.06.009] [PMID: 26142753]

Koelmans, A.A., Besseling, E., Wegner, A., Foekema, E.M. (2013). Plastic as a carrier of POPs to aquatic organisms: A model analysis. *Environ. Sci. Technol., 47*(14), 7812-7820.
[http://dx.doi.org/10.1021/es401169n] [PMID: 23758580]

Kim, D., Chae, Y., An, Y.J. (2017). Mixture toxicity of nickel and microplastics with different functional groups on *Daphnia magna. Environ. Sci. Technol., 51*(21), 12852-12858.
[http://dx.doi.org/10.1021/acs.est.7b03732] [PMID: 29019667]

Lei, L., Wu, S., Lu, S., Liu, M., Song, Y., Fu, Z., Shi, H., Raley-Susman, K.M., He, D. (2018). Microplastic particles cause intestinal damage and other adverse effects in zebrafish *Danio rerio* and nematode *Caenorhabditis elegans. Sci. Total Environ., 619-620*, 1-8.
[http://dx.doi.org/10.1016/j.scitotenv.2017.11.103] [PMID: 29136530]

Lehner, R., Weder, C., Petri-Fink, A., Rothen-Rutishauser, B. (2019). Emergence of nanoplastic in the environment and possible impact on human health. *Environ. Sci. Technol., 53*(4), 1748-1765.
[http://dx.doi.org/10.1021/acs.est.8b05512] [PMID: 30629421]

Lee, K.W., Shim, W.J., Kwon, O.Y., Kang, J.H. (2013). Size-dependent effects of micro polystyrene particles in the marine copepod *Tigriopus japonicus. Environ. Sci. Technol., 47*(19), 11278-11283.
[http://dx.doi.org/10.1021/es401932b] [PMID: 23988225]

Lithner, D. (2011). Environmental and health hazards of chemicals in plastic polymers and products. Gothenburg: University of Gothenburg.

Luís, L.G., Ferreira, P., Fonte, E., Oliveira, M., Guilhermino, L. (2015). Does the presence of microplastics influence the acute toxicity of chromium(VI) to early juveniles of the common goby (*Pomatoschistus microps*)? A study with juveniles from two wild estuarine populations. *Aquat. Toxicol., 164*, 163-174.
[http://dx.doi.org/10.1016/j.aquatox.2015.04.018] [PMID: 26004740]

Lusher, A., Hollman, P., Mendoza-Hill, J. (2017). Microplastics in fisheries and aquaculture: Status of knowledge on their occurrence and implications for aquatic organisms and food safety. *FAO Fish. Aquac. Tech. Pap., 615*.

Lu, Y., Zhang, Y., Deng, Y., Jiang, W., Zhao, Y., Geng, J., Ding, L., Ren, H. (2016). Uptake and accumulation of polystyrene microplastics in zebrafish (*Danio rerio*) and toxic effects in liver. *Environ. Sci. Technol., 50*(7), 4054-4060.
[http://dx.doi.org/10.1021/acs.est.6b00183] [PMID: 26950772]

Mato, Y., Isobe, T., Takada, H., Kanehiro, H., Ohtake, C., Kaminuma, T. (2001). Plastic resin pellets as a transport medium for toxic chemicals in the marine environment. *Environ. Sci. Technol., 35*(2), 318-324.
[http://dx.doi.org/10.1021/es0010498] [PMID: 11347604]

Mattsson, K., Johnson, E.V., Malmendal, A., Linse, S., Hansson, L.A., Cedervall, T. (2017). Brain damage and behavioural disorders in fish induced by plastic nanoparticles delivered through the food chain. *Sci. Rep., 7*(1), 11452.
[http://dx.doi.org/10.1038/s41598-017-10813-0] [PMID: 28904346]

Ma, Y., Huang, A., Cao, S., Sun, F., Wang, L., Guo, H., Ji, R. (2016). Effects of nanoplastics and microplastics on toxicity, bioaccumulation, and environmental fate of phenanthrene in fresh water. *Environ. Pollut., 219*, 166-173.
[http://dx.doi.org/10.1016/j.envpol.2016.10.061] [PMID: 27814532]

Murray, F., Cowie, P.R. (2011). Plastic contamination in the decapod crustacean Nephrops norvegicus (Linnaeus, 1758). *Mar. Pollut. Bull., 62*(6), 1207-1217.
[http://dx.doi.org/10.1016/j.marpolbul.2011.03.032] [PMID: 21497854]

Ng, E.L., Huerta Lwanga, E., Eldridge, S.M., Johnston, P., Hu, H.W., Geissen, V., Chen, D. (2018). An overview of microplastic and nanoplastic pollution in agroecosystems. *Sci. Total Environ., 627*, 1377-1388.
[http://dx.doi.org/10.1016/j.scitotenv.2018.01.341] [PMID: 30857101]

Oliveira, M., Ribeiro, A., Hylland, K., Guilhermino, L. (2013). Single and combined effects of microplastics and pyrene on juveniles (0+ group) of the common goby Pomatoschistus microps (Teleostei, Gobiidae). Ecol. Indic., 34, 641-647..
[http://dx.doi.org/10.1016/j.envpol.2014.10.026] [PMID: 25463733]

Ogonowski, M., Schür, C., Jarsén, Å., Gorokhova, E. (2016). The effects of natural and anthropogenic microparticles on individual fitness in *Daphnia magna*. *PLoS One, 11*(5), e0155063.
[http://dx.doi.org/10.1371/journal.pone.0155063] [PMID: 27176452]

Oliveira, M., Ribeiro, A., Hylland, K., Guilhermino, L. (2013). Single and combined effects of microplastics and pyrene on juveniles (0+ group) of the common goby Pomatoschistus microps (Teleostei, Gobiidae). *Ecol. Indic., 34*, 641-647.
[http://dx.doi.org/10.1016/j.ecolind.2013.06.019]

O'Donovan, S., Mestre, N.C., Abel, S., Fonseca, T.G., Carteny, C.C., Cormier, B., Keiter, S.H., Bebianno, M.J., Donovan, S.O. (2018). Ecotoxicological effects of chemical contaminants adsorbed to microplastics in the clam scrobicularia plana. *Front. Mar. Sci., 5*, 143.
[http://dx.doi.org/10.3389/fmars.2018.00143]

Phuong, N.N., Zalouk-Vergnoux, A., Poirier, L., Kamari, A., Châtel, A., Mouneyrac, C., Lagarde, F. (2016). Is there any consistency between the microplastics found in the field and those used in laboratory experiments? *Environ. Pollut., 211*, 111-123.
[http://dx.doi.org/10.1016/j.envpol.2015.12.035] [PMID: 26745396]

Rehse, S., Kloas, W., Zarfl, C. (2018). Microplastics reduce short-term effects of environmental contaminants. Part I: Effects of bisphenol a on freshwater zooplankton are lower in presence of polyamide particles. *Int. J. Environ. Res. Public Health, 15*(2), 280.
[http://dx.doi.org/10.3390/ijerph15020280] [PMID: 29415519]

Ribeiro, F., Garcia, A.R., Pereira, B.P., Fonseca, M., Mestre, N.C., Fonseca, T.G., Ilharco, L.M., Bebianno, M.J., Ilharco, L.M., Bebianno, M.J. (2017). Microplastics effects in *Scrobicularia plana*. *Mar. Pollut. Bull., 122*(1-2), 379-391.
[http://dx.doi.org/10.1016/j.marpolbul.2017.06.078] [PMID: 28684108]

Rochman, C.M., Hoh, E., Hentschel, B.T., Kaye, S. (2013). Long-term field measurement of sorption of organic contaminants to five types of plastic pellets: Implications for plastic marine debris. *Environ. Sci. Technol., 47*(3).
[http://dx.doi.org/10.1021/es303700s] [PMID: 23270427]

Savoca, S., Capillo, G., Mancuso, M., Faggio, C., Panarello, G., Crupi, R., Bonsignore, M., D'Urso, L., Compagnini, G., Neri, F., Fazio, E., Romeo, T., Bottari, T., Spanò, N. (2019). Detection of artificial cellulose microfibers in Boops boops from the northern coasts of Sicily (Central Mediterranean). *Sci. Total. Environ., 691*, 455-465.
[http://dx.doi.org/10.1016/j.scitotenv.2019.07.148] [PMID: 31323590]

Savoca, S, Capillo, G, Mancuso, M, Bottari, T, Crupi, R, Branca, C, Romano, V, Faggio, C, D'angelo, G, Spanò, N (2019). Microplastics occurrence in the tyrrhenian waters and in the gastrointestinal tract of two congener species of seabreams. *Environ. Toxicol. Pharmacol., 67*, 35-41.

Smith, M., Love, D.C., Rochman, C.M., Neff, R.A. (2018). Microplastics in seafood and the implications for human health. *Curr. Environ. Health Rep., 5*(3), 375-386.
[http://dx.doi.org/10.1007/s40572-018-0206-z] [PMID: 30116998]

Strungaru, S.A., Jijie, R., Nicoara, M., Plavan, G., Faggio, C. (2019). Micro- (nano) plastics in freshwater ecosystems: Abundance, toxicological impact and quantification methodology. *Trends Analyt. Chem., 110*, 116-128.
[http://dx.doi.org/10.1016/j.trac.2018.10.025]

Stara, A., Bellinvia, R., Velisek, J., Strouhova, A., Kouba, A., Faggio, C. (2019). Acute exposure of neonicotinoid pesticide on common yabby (*Cherax destructor*). *Sci. Total Environ., 665*, 718-723.

[http://dx.doi.org/10.1016/j.scitotenv.2019.02.202] [PMID: 30780017]

Stara, A., Pagano, M., Capillo, G., Fabrello, J., Sandova, M., Vazzana, I., Zuskova, E., Velisek, J., Matozzo, V., Faggio, C. (2020). Assessing the effects of neonicotinoid insecticide on the bivalve mollusc *Mytilus galloprovincialis. Sci. Total Environ., 700*, 134914.
[http://dx.doi.org/10.1016/j.scitotenv.2019.134914] [PMID: 31706094]

Sleight, V.A., Bakir, A., Thompson, R.C., Henry, T.B. (2017). Assessment of microplastic-sorbed contaminant bioavailability through analysis of biomarker gene expression in larval zebrafish. *Mar. Pollut. Bull., 116*(1-2), 291-297.
[http://dx.doi.org/10.1016/j.marpolbul.2016.12.055] [PMID: 28089550]

US EPA. (2018). *Advancing Sustainable Materials Management 2015 Tables and Figures: Assessing Trends in Material Generation.* Recycling, Composting, Combustion with Energy Recovery and Landfilling in the United States.

Van Cauwenberghe, L., Janssen, C.R. (2014). Microplastics in bivalves cultured for human consumption. *Environ. Pollut., 193*, 65-70.
[http://dx.doi.org/10.1016/j.envpol.2014.06.010] [PMID: 25005888]

Van Cauwenberghe, L., Claessens, M., Vandegehuchte, M.B., Janssen, C.R. (2015). Microplastics are taken up by mussels (*Mytilus edulis*) and lugworms (*Arenicola marina*) living in natural habitats. *Environ. Pollut., 199*, 10-17.
[http://dx.doi.org/10.1016/j.envpol.2015.01.008] [PMID: 25617854]

Veneman, W.J., Spaink, H.P., Brun, N.R., Bosker, T., Vijver, M.G. (2017). Pathway analysis of systemic transcriptome responses to injected polystyrene particles in zebrafish larvae. *Aquat. Toxicol., 190*, 112-120.
[http://dx.doi.org/10.1016/j.aquatox.2017.06.014] [PMID: 28704660]

von Moos, N., Burkhardt-Holm, P., Köhler, A. (2012). Uptake and effects of microplastics on cells and tissue of the blue mussel *Mytilus edulis* L. after an experimental exposure. *Environ. Sci. Technol., 46*(20), 11327-11335.
[http://dx.doi.org/10.1021/es302332w] [PMID: 22963286]

Wagner, M., Scherer, C., Alvarez-Muñoz, D., Brennholt, N., Bourrain, X., Buchinger, S., Fries, E., Grosbois, C., Klasmeier, J., Marti, T., Rodriguez-Mozaz, S., Urbatzka, R., Vethaak, A.D., Winther-Nielsen, M., Reifferscheid, G. (2014). Microplastics in freshwater ecosystems: What we know and what we need to know. *Environ. Sci. Eur., 26*(1), 12.
[http://dx.doi.org/10.1186/s12302-014-0012-7] [PMID: 28936382]

Watts, A.J.R., Lewis, C., Goodhead, R.M., Beckett, S.J., Moger, J., Tyler, C.R., Galloway, T.S. (2014). Uptake and retention of microplastics by the shore crab *Carcinus maenas. Environ. Sci. Technol., 48*(15), 8823-8830.
[http://dx.doi.org/10.1021/es501090e] [PMID: 24972075]

WHO. (2010). Preventing Disease through Healthy Environments. Exposure to Cadmium: A Major Public Health Concern. Geneva, Switzerland: World Health Organization.

Weis, J., Andrews, C.J., Dyksen, J.E. (2015). Human health impact of microplastics and nanoplastics. NJDEP - Science Advisory Board.

Wieczorek, A.M., Morrison, L., Croot, P.L., Allcock, A.L., MacLoughlin, E., Savard, O., Brownlow, H., Doyle, T.K. (2018). Frequency of microplastics in mesopelagic fishes from the northwest atlantic. *Front. Mar. Sci., 5*, 39.
[http://dx.doi.org/10.3389/fmars.2018.00039]

Wright, S.L., Thompson, R.C., Galloway, T.S. (2013). The physical impacts of microplastics on marine organisms: A review. *Environ. Pollut., 178*, 483-492.
[http://dx.doi.org/10.1016/j.envpol.2013.02.031] [PMID: 23545014]

Overview of Marine Plastic Pollution in the Moroccan Mediterranean

Bilal Mghili[1,*], Mohamed Analla[1] and **Mustapha Aksissou[1]**

[1] *LESCB, URL-CNRST N° 18, Abdelmalek Essaadi University, Faculty of Sciences, Tetouan, Morocco*

Abstract: Plastic debris has become the main component of marine litter in the Moroccan Mediterranean due to the massive consumption of plastic and poor plastic waste management. In Morocco, plastic pollution has been a subject of increasing environmental concern in the last few years. This literature review was conducted to collect current data on plastic pollution in the Moroccan Mediterranean, considering the presence of marine debris as well as macroplastics and microplastics in different compartments. Our study shows that, until now, very few studies have been carried out and there is a lack of information, especially on the prevalence of plastic debris in the water environment, sea floor, and aquatic animals. In general, plastic is the most predominant waste on the beaches of the Moroccan Mediterranean, always contributing to more than 50% of the total composition of the waste encountered. Based on the records, tourism, recreational activities, and fishing are one of the main sources of plastic accumulation in the Moroccan Mediterranean. This was due to a lack of awareness among beach users. Awareness and behavior change is key to minimizing plastic waste on Morocco's beaches and coasts. In addition, all aspects of waste management must be improved. The beaches of the Moroccan Mediterranean have also been contaminated by microplastics. A significant positive correlation was also observed between human population density and industrial activity on microplastic abundance. Microplastic has only been found in a few commercial fish species and sea turtles, but more work will be needed in the future.

Keywords: Beach, Characterization, Coast, Debris, Ingestion, Macroplastics, Management, Marine debris, Marine fauna, Marine litter, Mediterranean, Microplastics, Mismanagement, Mitigation, Morocco, Pollution, Quantification, Recycling.

INTRODUCTION

Plastic pollution of the oceans has been identified as one of the major environmental challenges of the 21st century (De-la-Torre *et al.*, 2021). Each year,

---

* **Corresponding author Bilal Mghili:** LESCB, URL-CNRST N° 18, Abdelmalek Essaadi University, Faculty of Sciences, Tetouan, Morocco; Tel: +212610103615; E-mail: b.mghili@uae.ac.ma

millions of tons of plastic litter are released into the sea with detrimental environmental implications. Tourism and recreation activities, agriculture, and poor management of municipal and industrial solid waste, and dumping of poorly treated/untreated wastewater are considered the major land-based sources of marine debris, while marine sources include fisheries, shipping, and aquaculture (Li *et al.*, 2016).

The Mediterranean Sea has been recognized as one of the most affected seas by plastic pollution in the world (Suaria *et al.*, 2016; Fossi *et al.*, 2018; Constantino *et al.*, 2019). The coasts of the Mediterranean are inhabited by about 10% of the global coastal population and the basin constitutes one of the world's major shipping routes. It receives water from densely populated basins. During the last decades, large research activity has been conducted on the study of plastic pollution in the Mediterranean Sea. These studies indicate the presence, quantity, sources, and to some degree, the impacts of marine debris on beaches (Vlachogianni, 2019; Mghili *et al.*, 2020), the seafloor (Fortibuoni *et al.*, 2019), the sea surface (Arcangeli *et al.*, 2018; Zeri *et al.*, 2018), and biota (Anastasopoulou *et al.*, 2018). In the Mediterranean, the ingestion of plastics by marine species is one of the most documented impacts (Fossi *et al.*, 2018). More than 40 articles on the incidence of marine debris ingestion in marine biota in the Mediterranean have been published in the last five years. Despite this mass of information, the overall knowledge of the plastic pollution issue in the Mediterranean Sea is still limited and fragmented. Most studies have been carried out in the western Mediterranean Sea, while the southwestern Mediterranean has been less studied.

The Moroccan Mediterranean coastline extends 512 km from Tangier in the west to Saidia in the east. It is recognized for its high marine biodiversity and a wide variety of coastal landscapes. In recent years, the population of the Moroccan Mediterranean coast has undergone a significant change. The population has doubled from 1.5 million in 1971 to more than 2.8 million in 2014 (RGPH, 2014). In parallel with the increase in the coastal population, economic activities such as industry, agriculture, fishing, aquaculture, and tourism are well developed. In the list of top 20 waste-producing countries (Jambeck *et al.*, 2015), Morocco is placed eighteenth on this list with 0.31 million metric tons of mismanaged plastic waste per year. According to the Ministry of Industry, 26 billion plastic bags are consumed annually in Morocco, or 800 bags per capita. The amount of plastic waste entering the sea from land in Morocco in 2010 was estimated to be between 0.05 and 0.12 million metric tons (Jambeck *et al.*, 2015). The majority of waste is deposited in landfills, while only 8% is recycled. Plastic pollution in the Moroccan Mediterranean has been demonstrated in several abiotic compartments, including beaches, sediments, sea floor, and biota. Despite this fact, information is still limited. There are still several gaps in knowledge about plastic pollution in the Moroccan Mediterranean.

The present study aimed to collect and analyze scientific papers that have studied plastic pollution in the Moroccan Mediterranean, to obtain an overview of the state of the art of studies carried out on plastic pollution in the Moroccan Mediterranean, and to define the possible gaps in this knowledge.

MATERIALS AND METHOD

Using the keywords "plastic", "marine litter", "marine debris", "marine environment", "coastal environments" and their respective French translations, all combined with "Morocco/Moroccan", we have performed a literature review in the search databases Google Scholar, Scopus and Web of Science. In this study, we retrieved book chapters and papers published in national and international journals. The conference papers were also included in this analysis as data because these papers are also considered important. Searches were also conducted on the references mentioned in the retrieved articles.

All articles (n = 18) treating plastic pollution in the Moroccan Mediterranean were included in this study. First, we analyze and report data on plastics on beaches, in the sediment, and on the seafloor. Then, we concentrate our attention on microplastics in the marine environment, including marine biota.

In this review, we aim to respond to the following questions: 1) How has scientific research on plastic pollution evolved in the Moroccan Mediterranean in recent years? 2) What are the themes covered? 3) What are the main types of debris in the Moroccan Mediterranean? 4) What are the sources of this debris 5) and What are the main gaps in scientific knowledge to guide future work?

RESULTS AND DISCUSSION

Scientific Production

Over the last few years, scientific production about marine debris has increased in the Moroccan Mediterranean (Fig. **1**). From this review, we retrieved 18 articles. Despite this, we have observed that the number of papers about plastic pollution in Morocco is low (X=3 papers/year). The total number of papers remains exceedingly low when compared to the global publication rate, reflecting a lack of studies on plastic pollution in the Moroccan Mediterranean.

Of the 18 studies identified in the Moroccan Mediterranean, the most common type of publication was scientific articles (n = 11; 61.13%), followed by conference papers (n = 5; 27.77%), reports (n = 1; 5.55%), and book chapters (n = 1; 5.55%). A large number of articles (28%) used the term "marine litter", but "marine debris", "litter" and "marine waste" were also used. Most of the recovered studies (77%) address the characterization and quantification of marine debris in coastal environments. Multiple environments were surveyed, and beaches were the most studied, with 12 publications. Sediments (4) and seabed (2) were the other environments studied. Four of the recovered works treated the subject of microplastics. The rest of the literature (3) focuses on the ingestion of MPs by marine fauna.

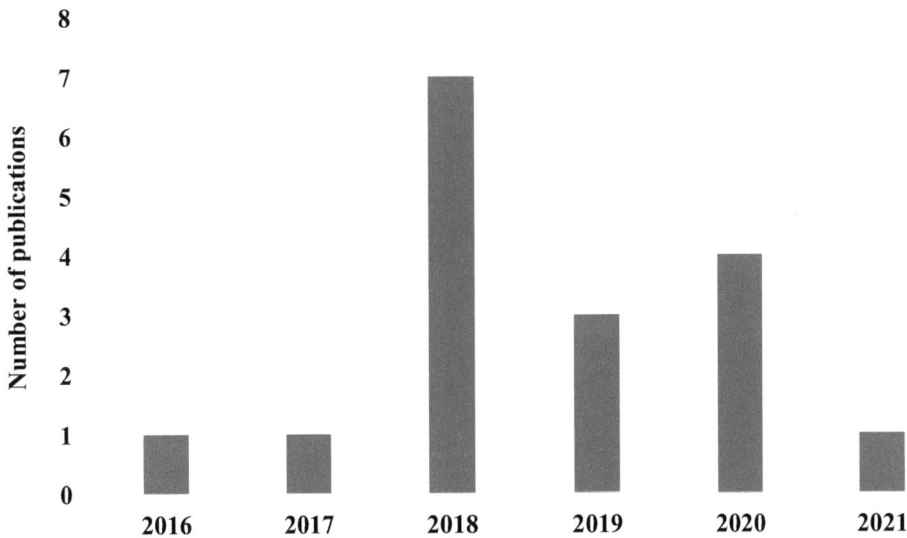

Fig. (1). Number of publications in the Moroccan Mediterranean retrieved by literature review.

Characterization and Quantification of Plastic Debris in Coastal and Marine Environments

Plastic Debris on Beaches

The majority of the recovered articles (66%) addressed this topic. Most of the studies on marine litter have examined the presence of these wastes on sandy beaches. Fig. (2) presents the geographical distribution of research about plastics along the Moroccan Mediterranean. The Tetouan coast has a large number of studies on the qualitative and quantitative evaluation of marine debris on its beaches. In the province of El-Hociema and the Nador-Oujda region, although it also has an extensive coastline, studies on the qualitative and quantitative evaluation of debris are very rare. Most studies assessed the presence of marine debris spatiotemporally, while 5 assessed only temporally and one only spatially. The principal methodology used for sampling was the perpendicular transect from the tide line to the vegetation. Transect widths ranged from 5 m to 20 m.

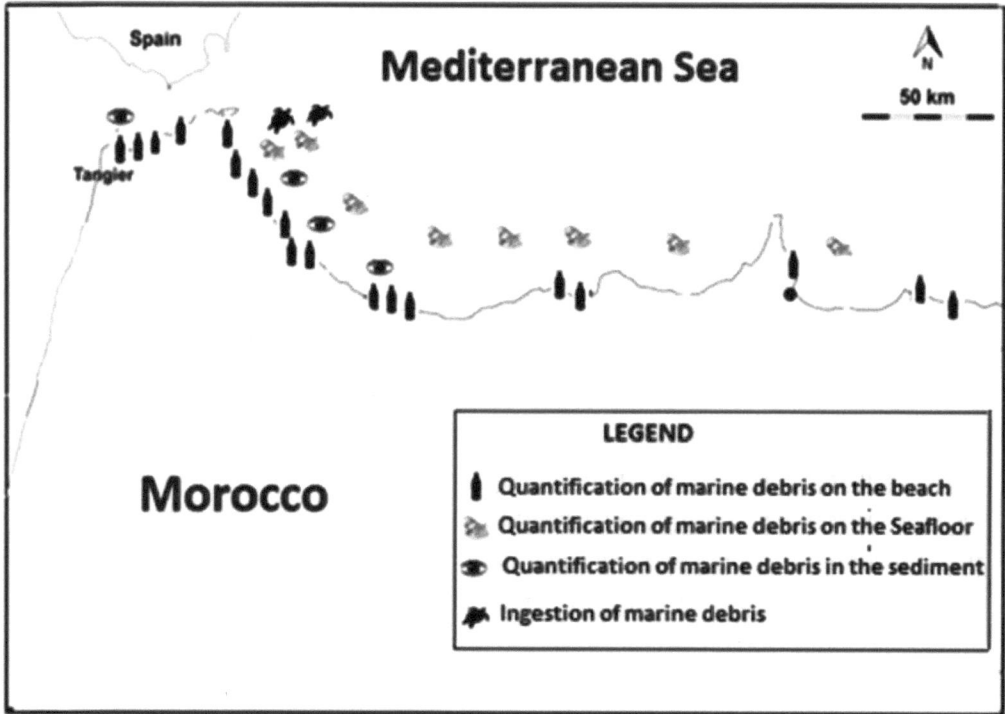

Fig. (2). Map showing the number of studies carried out on each theme, in the different coastal regions of the Moroccan Mediterranean.

The majority of works cited plastic as the predominant residue, always contributing more than 50% to the total composition of waste encountered (Table 1). This result is consistent with the work conducted around the world. The majority of the waste on Moroccan Mediterranean beaches is composed mainly of plastic with numerical percentages of 85% (Association Zero Zbel, 2018). The most visible types of plastic debris are plastic bags, chip packets and candy wrappers, caps and lids, food wrappers, and bottles. The seasonal variation shows that plastic is dominated in the four seasons on the beach of Martil and Oued Laou, with higher density in summer (Alshawafi *et al.*, 2016; Alshawafi *et al.*, 2017). On the beaches of Tangier classified as tourist, a percentage of 60% of litter was plastic (Alshawafi *et al.*, 2020), while on the nontourist beaches, this percentage was 86% (Alshawafi *et al.*, 2018a). Another study on the tourist beach of Tangier showed a large percentage of plastic (94%) (Nachite *et al.*, 2018a).

This difference can be explained by the sampling method used by the two studies. Recent surveys on the coast of Tetouan (Mghili *et al.*, 2020), and 14 beaches in the Moroccan Mediterranean (Maziane *et al.*, 2018; Nachite *et al.*, 2018b, 2019) did not show very different results, with a prevalence of plastic waste of 68% and 94% respectively. Vlachogianni *et al.* (2020) showed that plastics are generally the most important component of marine debris, sometimes constituting up to 62% of the debris. This phenomenon can also be observed on the seabed where 73% of the litter collected in benthic trawls is plastic (Loulad *et al.*, 2019). During the COVID-19 pandemic, personal protective equipment (PPE) became a common type of debris invading many different environments, including tourist beaches. Recently, Mghili *et al.* (2022) reported the occurrence of face masks on five beaches in the Moroccan Mediterranean. The presence of this type of PPE on the Moroccan beach threatens the marine environment with a new form of plastic pollution.

Table 1. Information on the main results of the work recovered on characterization and quantification of marine debris in the marine environment.

References	Sampled Environment	Analysis	Sampling Method	Total Items Collected	Waste Density	Prevalent Residue	% of Plastic	Unities
Alshawafi *et al.* (2016)	Beach	Temporal	Transect 5 m wide	11.80	NC	Plastic	53%	Kg
Alshawafi *et al.* (2017)	Beach	Temporal	Transect 5 m wide	12.87	NC	Plastic	57%	Kg
Alshawafi *et al.* (2018)	Beach	Temporal	Transect 5 m wide	6.90	NC	Plastic	86%	Kg
Zero Zbel Association (2018)	Beach	Spatio-temporel	Transect 400 m^2	36 280	NC	Plastic	85%	Items/m^2
Maziane *et al.* (2018)	Beach	Spatio-temporel	Transect from the low tide line to the base of the vegetation-20 m wide	8021	0.02–0.153	Plastic	94%	Items/m^2
Nachite *et al.* (2018a)	Beach	Spatio-temporel		12 207	0.06 ± 0.04	Plastic	67%	Items/m^2
Nachite *et al.* (2018b)	Beach	Temporal		3628	0.049 ±0.018	Plastic	94%	Items/m^2
Loulad *et al.* (2019)	Seafloor	Spatio-temporel	The total area surveyed was subdivided into five depth strata	497	26 ± 68 - 80 ±133	Plastic	73%	kg/km^2
Azaaouaj *et al.* (2019)	Seafloor	Spatial	Transects 100 m wide	881	0.997	Plastic	75%	Items/m^2

(Table 1) cont.....

References	Sampled Environment	Analysis	Sampling Method	Total Items Collected	Waste Density	Prevalent Residue	% of Plastic	Unities
Nachite *et al.* (2019)	Beach	Spatio-temporel	Transect from the low tide line to the base of the vegetation-20 m wide	21 943	0.054 ±0.036	Plastic	69%	Items/m^2
Mghili *et al.* (2020)	Beach	Spatio-temporel		7839	0.20 ± 0.098	Plastic	68%	Items/m^2
Vlachogianni *et al.* (2020)	Beach	Spatio-temporel		13 078	0.05–0.6	Plastic	62%	Items/m^2
Alshawafi *et al.* (2020)	Beach	Temporal	Transect 5 m wide	11.03	NC	Plastic	60%	Kg
Mghili *et al.* (2022)	Beach	Spatio-temporel	Transect 20 m wide	321	0.0011 ± 0.0007	Plastic (face masks)	97%	Items/m^2

The Abundance of Plastic Litter

The density of marine debris has been expressed differently in different studies. Some researchers use density as items/area (m^2) and others use only weight (kg) or several items. Lack of standardization causes difficulty in the comparison of results. In general, the mean litter densities reported on Moroccan beaches are similar and comparable to the values recorded by other studies conducted in the Mediterranean. For example, Vlachogianni *et al.* (2020) reported average litter densities ranging from 0.05 to 0.60 units/m^2. A similar density was reported by Nachite *et al.* (2018b, 2019) on fourteen beaches along the Moroccan Mediterranean. This average density was also observed on Tangier Beach by Nachite *et al.* (2018a). Maziane *et al.* (2018) confirmed that the density of marine litter on fourteen beaches is in the same order of magnitude as on other beaches in the Mediterranean. In comparison with previous studies, Mghili *et al.* (2020) reported a higher average density (0.20 m^{-2}).

The abundance of marine debris varies considerably from place to place, influenced by human activities, and meteorological, and hydrological conditions. The abundance of debris varies according to the typology of the beaches. For example, Mghili *et al.* (2020) reported that the dirtiest beaches are located in urbanized areas, while the cleanest beaches are observed in village areas. Also, Nachite *et al.* (2019) found that accumulated debris around the urbanized areas of the beach was very high, compared to the rural and village beaches. Similar results have been documented by Vlachogianni *et al.* (2020). On the Moroccan Mediterranean, urban beaches are used by the Moroccan population and tourists in winter, summer, and autumn, which is probably the source of debris accumulation. The village beaches are used for recreational activities in the summer. The variation in abundance between beaches is not only due to anthropogenic activity. For example, the sea surface has a significant influence on the location where litter and marine are accumulated. Some abundance of marine debris on beaches along the Moroccan Mediterranean has been attributed to exposure to currents, wind, and tide (Alshawafi *et al.*, 2017; Nachite *et al.*, 2018b). The proximity of rivers can also increase the abundance of waste on nearby beaches. For example, the input from the heavily contaminated Oued Laou and Martil rivers could be another reason for the accumulation of marine debris, especially in winter.

The beach cleanliness was evaluated through the Clean Coast Index (CCI) (Alkalay *et al.*, 2007). In different studies, the numbers found are generally low and the cleanliness of these beaches varies from very clean to moderately clean. Nachite *et al.* (2019) calculated the index on 14 beaches. According to the index: 87.5% of them are in the "Very Clean" category and only 12.5% in the "Clean" category. The highest CCI value was observed in autumn and the lowest value was in spring. The CCI attained higher values at resorts and rural zones. Moreover, Mghili *et al.* (2020) evaluated the cleanliness of three urban beaches and two village beaches. The majority of beaches are classified as clean. Of the five beaches, the lower CCI value was observed in autumn and spring, while higher CCI values were recorded in winter and summer. Vlachogianni *et al.* (2020) classified the majority of beaches as "very clean" according to the Clean Coast Index.

The abundance of debris on beaches varies between the four seasons of the year. On the coast of the Tetouan region, Mghili *et al.* (2020) noted a seasonal influence on the quantity of litter present in the sands, with the greatest quantities occurring during the summer season. In the summer, when the number of tourists is higher, a greater total amount of litter in units was observed. Alshawafi *et al.* (2016, 2017) collected marine debris on Oued Laou and Martil beaches over the four seasons and observed a large seasonal influence on the waste density found, with higher rates in winter. In winter, tourism is very limited, but precipitation and high tides can transport marine debris. It has also been established that there is a significant influence of wind that promotes debris abundance on beaches (Nachite *et al.*, 2018b). In the Aouchtam resort beach, the authors observed temporal differences, with a greater accumulation of waste in the autumn period (Alshawafi *et al.*, 2018a). In the Moroccan Mediterranean, the lowest concentration of debris was reported in the spring which is due to the low number of visitors (Nachite *et al.*, 2019; Mghili *et al.*, 2020). The amount of waste accumulated on the beaches depends on the location, season of the year, characteristics of the area, and precipitation.

Plastic Accumulation Source

About 57% of the studies specified the source of debris. Based on the records, tourism, recreational activities, and fishing are one of the main sources of plastic accumulation in the Moroccan Mediterranean. In general, the data collected on the beach highlighted that debris from land-based sources was more dominant (Alshawafi *et al.*, 2017). Several papers have documented the abundance of plastic on various beaches as a result of anthropogenic activities such as recreation and tourism. These studies suggest a strong relationship between debris abundance and beach use rates. Nachite *et al.* (2019), covering the entire Mediterranean, tested on 14 beaches whether the pollution by waste could be correlated with the presence of beach users. Nachite *et al.* (2019) showed that beach visitor density plays an important role in marine pollution. Vlachogianni (2020) considers recreational activities and coastal tourism, as well as poor solid waste management, as the main causes of waste accumulation on the coastline (82%). Moreover, Mghili *et al.* (2020) showed differences between the sampled points in five beaches in the Tetouan region, with more waste in the area where the number of users was higher. According to the same authors, the main source of marine debris was coastal/recreational activities (70.13%). In different studies, the values obtained are much higher than the Mediterranean average of 52% (UNEP/MAPMEDPOL, 2011) and the global average of 68.2% (Ocean Conservancy, 2011).

Land-based sources can also include solid waste disposal activities and landfills associated with household and industrial waste (Nachite *et al.*, 2019; Mghili *et al.*, 2020). Sewage stormwater discharges and rivers are important routes through which land-based waste is transported.

Fishing is also known to generate large amounts of marine debris (Bouzekry *et al.*, 2022). Fishing is an important source in this area; being associated with 3% of the elements reported on the beaches studied (Mghili *et al.*, 2020). . A similar percentage was observed by Nachite *et al.* (2019) and Vlachogianni (2020) along the Moroccan Mediterranean. Shipping and maritime activities are also other sources of plastic accumulation. Vlachogianni (2020) observed that the levels of shipping-related debris on the coast of Morocco are significantly lower.

Plastic Pollution on the Seafloor

The seabed has also been studied. Loulad *et al.* (2019) studied the abundance, composition, and distribution of seabed litter in the Moroccan Mediterranean Sea. The average abundance of waste on the seafloor differed from survey to survey, generally ranging from 26 ± 68 Kg/km^2 to 80 ± 133 Kg/km^2. Plastic materials reached 73% of the total debris catch, which represented a mean abundance of 25.53 ± 58.79 kg/km^2, followed by rubber (12%), textile/clothing (8%), metal (3%), glass/ceramic (0.3%), and unidentified materials (2.7%). The majority of the plastic items were plastic bags, plastic bottles, shoes, gloves, and octopus pots. Analysis of the data indicates that the abundance and distribution of marine debris were strongly affected by local anthropogenic activities and river inputs.

During a litter fishing campaign, Azaaouaj and Nachite (2019) assessed the presence of marine debris on the seabed in the Fnideq area. Three sites were sampled at 2.5-3.5 m depth, with the participation of 4 artisanal fishing boats and 6 divers. A total of 881 items were collected with a weight of 68943 kg. The density of benthic debris at Fnideq is 0.937 items/m^2. Plastics represented about 75% of the collected waste.

Alshawafi *et al.* (2018b) evaluated the presence of debris on fishing nets. Out of thirty fishing operations, twenty-six operations have revealed the presence of marine debris in the fishing net. These studies show that our marine environments are highly polluted by plastics. This hypothesis is highly supported by the waste observed in fishing boats in their nets during the fishing process.

Microplastic Pollution in Sediments

There are a few published articles demonstrating the presence of microplastic (MP) in sediments. Sampling was conducted in a completely random manner within a few 5 m transects. Portions of sediment were collected from a 50×50 cm square at a depth of approximately 5 cm. In almost all studies, identification was performed by light microscopy. As can be seen, the study focused on four distinct regions: two urban beaches (Tangier and Martil) and a beach resort of Aouchtam, and a beach village Oued Laou. There is no record of studies conducted in the Nador-Oujda region with MP.

Although there are many studies focused on the identification and characterization of MP, standardized methods are lacking. In this work, the degree of pollution of the sediment by MP was expressed in weight (g). Alshawafi *et al.* (2016, 2017, 2018a, 2020) presented the presence of MPs in the coastal sediments of the Tetouan region. The maximum number of MPs was recorded on the urban beaches of Tangier and Martil and the minimum number was on the beaches of Oued Laou and Achoutam. A significant positive correlation was also observed between human population density and industrial activity on MP abundance. In general, the density of PM detected is higher in winter than in other periods in all years considered. In these four studies, the main forms of MP were fragments, foams, films, lines, and fibers.

PLASTIC POLLUTION IN BIOTA

In this review, this topic was the least studied by researchers. The studies quantifying the ingestion of MPs are concentrated in the Tetouan region. Three studies have reported the ingestion of debris by fish and sea turtles. Microplastic is often mistaken for food, which has an impact on the organisms that ingest it. Alshawafi *et al.* (2018b) analyzed a total of 30 individuals of *Diplodus cervinus* for MP ingestion in the Tetouan region. Results show that 20% of individuals ingested MPs. The same authors analyzed the presence of PM in the gastrointestinal contents of captured *Auxis thazard*. Of the 30 fish gastrointestinal tracts examined, 10% contained some type of PM. The microplastics present in the stomachs of both fish were a collection of fragments, pellets, and lines ranging in size from 0.5 to 4 mm. Alshawafi *et al.* (2018b) demonstrated that benthic fish had higher MPs, supporting the hypothesis on the relationship between organism habitats and feeding behavior. Also, Ahechti *et al.* (2018) evaluated the presence of MPs in four commercial species: *Boops boops*, *Scomber, scombrus*, *Mugil cephalus,* and *Sardina pilchardus*. These authors reported the presence of MPs in the stomachs of all four fish.

In the Moroccan Mediterranean, two species of marine turtles, the loggerhead turtle, *Caretta caretta*, and the leatherback turtle, *Dermochelys coriacea*, have been present and both are on the IUCN Red List. The study by Mghili *et al.* (2023b) was the first and only record of marine debris in the digestive tract of sea turtles in the Moroccan Mediterranean. The digestive tract contents of 20 loggerhead turtles (*C. caretta*) and 3 leatherbacks (*D. coriacea*) were examined for marine debris. Marine debris was present in the digestive tract of 80% of loggerheads and 33% of leatherbacks. The ingested debris consisted mainly of fishing hooks and nets (62.5%) and wood fragments (12.5%). The plastic ingested by loggerhead turtles constituted about 5% inorganic debris.

STATUS OF PLASTIC MANAGEMENT AND MITIGATION IN MOROCCO

Plastic management is a complex question that engages national policy, societal behavior, and cross-border collaboration. In Morocco, the issue of plastic mitigation has attracted the attention of a diverse range of actors including government, academia, and civil society organizations over the past few years. There are no publications on the management of plastics in the marine environment. Nachite *et al.* (2018b) discussed the policies for marine litter management in Morocco. There is no quantified study of the different sources of plastic leakage at the national level.

In response to the environmental risks associated with poor waste management, Morocco has implemented a series of strategic actions to reform the waste management sector. The Kingdom of Morocco has established an integrated approach that includes a legal and institutional framework, financial resources, and strategic initiatives. It has established a collection, treatment, storage, sorting, and disposal system for waste.

First, Morocco has reinforced its environmental policy. Morocco has signed many international conventions for the protection of the marine environment and several regions for the protection of the Atlantic and Mediterranean Seas. In 1972, Morocco signed the London "Convention on the Prevention of Marine Pollution by Dumping of Wastes and Other Matters". It has also signed the "Basel Convention on the Control of Transboundary Movements of Hazardous Wastes and their Disposal" the "International Convention for the Prevention of Pollution from Ships (MARPOL 73/78) in 1989 and the "Barcelona Convention for the Protection of the Marine Environment and the Coastal Region of the Mediterranean" of 1976.

In addition to improving policy environments, the Moroccan government has approved several laws to combat plastic pollution. The most important laws are Act No. 11-03 (2003) for the protection and conservation of the environment, which established a variety of environmental policy measures; Act No. 12-03 (2003) on environmental impact assessment, and Act No. 28-00 (2006) on waste management and disposal, which aims to reduce the harmful impacts of waste on human health and the environment. In 2015, Morocco adopted Act No. 81-12 (2015) on the coastline, which aims to prevent and mitigate pollution and degradation of the coastline through the rehabilitation of polluted and damaged regions.

Recently, Morocco has also adopted a law in the fight against plastic pollution: since July 1, 2016, Act No. 77-15 prohibits the production, import, sale, or distribution of single-use plastic bags. Five years after the ban on plastic bags in Morocco, they are still widely used. There is a need to take even more consistent action against plastic waste, including stricter enforcement of the Zero Mika law.

These international engagements of Morocco have been followed by the development of numerous national strategies and programs:

The National Municipal Solid Waste Management program (2008-2022) seeks to improve the management of municipal solid waste by assuring the collection of 90% of the waste to realize landfills and recovery centers for all urban centers (100%), and by developing the sorting-recycling-recovery chain to achieve the recycling of 20% of the waste by 2021.

The National Program for the Collection and Elimination of Plastic Bags aimed to conduct campaigns for the disposal of plastic bags and educate citizens on the use of alternative products. The program also aimed to mobilize local actors and civil society for collection and disposal. The program was conducted in 2011-2012 and resulted in the collection and disposal of 1,000 tons of disposed plastic bags throughout the country.

The National Program for the Prevention and Fight against Industrial Pollution was initiated in 2009 to reduce the production of pollutants and waste from industrial production.

To minimize plastic pollution, the initiatives "Clean Beaches" and "Blue Flag", were initiated, respectively, in 1999 and 2002 by the Foundation Mohammed VI for Environmental Protection. These two initiatives aimed to clean the beaches and transmit real environmental education to the summer people.

Plastic pollution is a global environmental problem and international collaboration will be needed to find effective solutions. Within the framework of the "Association Agreement", Morocco and the European Union are actively cooperating, among others, on the monitoring and prevention of marine pollution.

RECOMMENDATIONS AND FUTURE STUDIES

Despite the low number of papers found, we can say that research on marine plastic has been increasing in Morocco. In terms of localization, most publications deal with the state of marine debris in the Tetouan region. It is noted that only three publications deal with the state of marine debris in the region of El Hociema and Nador-Oujda, they have been published. However, it is most important to continue encouraging debris plastic research, particularly in locations where there is no data on the subject.

There are enormous gaps in comparability between articles due to the lack of a standard methodology for sampling. Different units were used to quantify debris such as weight, density, and percentage. Therefore, it is difficult to compare the available results between studies. From a global point of view, it is necessary to standardize protocols for normalizing the findings obtained by different methods to compare the data collected around the world. Looking only at data from beaches reported with the same unit of measurement (items/m^2) between 2018 and 2020, it seems that the situation is almost stable.

In the Moroccan Mediterranean, information is lacking on numerous aspects of plastic pollution. The review showed that several compartments have been studied, but sandy beaches are the most investigated ecosystems, possibly because it is easier to access and to conduct sampling. All marine ecosystems, including coral reefs, seagrass beds, and benthic systems must be monitored. This is necessary to understand the extent of plastic pollution and its impacts. Regarding MPs, there are no complete studies. To date, there have been no studies on the presence of MPs in the water column and on the seabed. Toxicological studies on MPs, especially their possible synergistic effects with other contaminants, also need to be of concern. Only one study has evaluated the adsorption capacity of MPs for different metals according to the physicochemical conditions of aquatic environments (Ahechti *et al.*, 2020). Studies on nanoplastics are also lacking.

The Moroccan Mediterranean has enormous biodiversity that is threatened by the presence of this debris. Overall, few studies examined the ingestion of MP by marine biota in the Moroccan Mediterranean. There is no published information regarding the ingestion of MP by marine mammals, birds, and invertebrates. Studies have documented that a wide variety of commercially important fish species are polluted with MPs in their stomachs. The results of these studies highlight the potential danger of absorption of MPs by the human system through direct consumption of these fish. Discarded marine debris is one of the potential vectors of marine invasive species. There are no publications yet dealing with the presence of invasive organisms on plastic in Morocco.

Factors influencing the accumulation of debris in coastal areas include wind, tide, and currents. Little is known about the trends in the role of these factors in debris accumulation. The Moroccan Mediterranean receives discharges from several rivers. Very little data exist for the assessment of marine debris from rivers. In the Moroccan Mediterranean, rivers are potentially important pathways into the marine environment. It is therefore important to quantify the extent and types of litter entering the sea from rivers.

Poor solid waste management is also one of the major environmental problems in the Moroccan Mediterranean and a probable source of marine waste. The scientific articles mentioned the acts and legislative measures currently used to address plastic pollution. In Morocco, there is no published article on the effectiveness of laws, policies, and governance to control waste flows. Waste leaks that enter the marine ecosystem need to be thoroughly investigated as we need to identify the potential source of leaks and possible preventive measures. As beach users are the main sources of marine debris, education on the impacts of plastic debris is needed to increase public awareness and build self-responsibility among the public. Awareness must also be supported by the establishment of a waste disposal infrastructure. Assessment of environmental perception and environmental education are important to determine the level of society's knowledge of the marine debris problem and to assess methods for raising awareness among uninformed people.

During the pandemic, face masks have become a common type of waste on the beaches of the Moroccan Mediterranean. This new type of pollution will aggravate plastic pollution in the Mediterranean and pose a threat to marine biota. Monitoring of PPE items related to the COVID-19 pandemic in the marine environment is necessary to better understand the degree of impact of this waste.

CONCLUSION

The abundance and distribution of marine litter on some Moroccan Mediterranean beaches were only reported in the last year. The present study suggests that plastic debris is present in several compartments. This marine debris comes mainly from tourism and fishing. Ingestion of anthropogenic debris by marine animals has been documented in a number of species in the Moroccan Mediterranean. In Morocco, there are appropriate laws, policies, and governance frameworks to address the challenge of plastic waste, but all aspects of waste management need to be improved. It is necessary to evaluate the effectiveness of the actions adopted for the prevention, mitigation, and control of marine litter, and the surveillance and enforcement of laws and policies.

REFERENCES

Ahechti, M., Benomar, M., El Alami, M., Mendiguchía, C. (2020). Metal adsorption by microplastics in aquatic environments under controlled conditions: Exposure time, pH and salinity. *Int. J. Environ. Anal. Chem.*, 1-8.

Ahechti, M., Benomar, M., El Alami, M., Mendiguchía, C. (2018). Microplastics and metal in commercial fish species from the Moroccan coast. *VI International Symposium on Marine Sciences Vigo (Spain).*

Alkalay, R., Pasternak, G., Zask, A. (2007). Clean-coast index—A new approach for beach cleanliness assessment. *Ocean Coast. Manage.*, *50*(5-6), 352-362.
[http://dx.doi.org/10.1016/j.ocecoaman.2006.10.002]

Alshawafi, A., Analla, M., Alwashali, E., Kassout, J., Aksissou, M. (2015). Presence the marine waste on the natural coast: Aouchtam, Morocco. *Int. J. Sci. Eng. Res.*, *9*, 8.

Alshawafi, A., Analla, M., Alwashali, E., Aksissou, M. (2020). Seasonal variation of marine litter in tangier coast: Quantitative and classificative study. *Open. J. Environ. Biol.*, *5*(1), 007-013.

Alshawafi, A., Analla, M., Alwashali, E., Aksissou, M. (2017). Assessment of marine debris on the coastal wetland of martil in the north-east of morocco. *Mar. Pollut. Bull.*, *117*(1-2), 302-310.
[http://dx.doi.org/10.1016/j.marpolbul.2017.01.079] [PMID: 28189368]

Alshawafi, A., Analla, M., Aksissou, M. (2016). Monitoring the marine debris at coastal fish oued laou on mediterranean sea, morocco. *Eur. J. Sci. Res.*, *143*, 26-35.

Alshawafi, A., Analla, M., Alwashali, E., Ahechti, M., Aksissou, M. (2018). Impacts of marine waste, ingestion of microplastic in the fish, impact on fishing yield, m'diq, morocco. *Int. J. Mar. Biol. Res.*, *3*(2), 1-14.
[http://dx.doi.org/10.15226/24754706/3/2/00125]

Anastasopoulou, A., Kovač Viršek, M., Bojanić Varezić, D., Digka, N., Fortibuoni, T., Koren, S., Mandić,

M., Mytilineou, C., Pešić, A., Ronchi, F., Šiljić, J., Torre, M., Tsangaris, C., Tutman, P. (2018). Assessment on marine litter ingested by fish in the Adriatic and NE Ionian Sea macro-region (Mediterranean). *Mar. Pollut. Bull., 133*, 841e851.

Arcangeli, A., Campana, I., Angeletti, D., Atzori, F., Azzolin, M., Carosso, L., Di Miccoli, V., Giacoletti, A., Gregorietti, M., Luperini, C., Paraboschi, M., Pellegrino, G., Ramazio, M., Sarà, G., Crosti, R. (2018). Amount, composition, and spatial distribution of floating macro litter along fixed trans-border transects in the Mediterranean basin. *Mar. Pollut. Bull., 129*(2), 545-554.
[http://dx.doi.org/10.1016/j.marpolbul.2017.10.028]

Association Zero Zbel. Analyse des déchets sur les plages marocaines. Available from: http://bit.ly/DechetsPlagesMaroc (accessed 2021-07-25)

Azaaouaj, S., Nachite, D. (2019). Fishing for litter at the port of Fnideq (NW Morocco). *II Congreso de Jóvenes Investigadonos del Mar Malaga.*

Benhardouze, W., Aksissou, M., Tiwari, M. (2015). Impact of pollution on marine turtles (*C. caretta* and *D. coreacea*) in NW OF Morocco. *35ᵗʰ Annual Symposium on Sea Turtle Biology and Conservation.* Dalaman, Turkey.

Constantino, E., Martins, I., Salazar Sierra, J.M., Bessa, F. (2019). Abundance and composition of floating marine macro litter on the eastern sector of the Mediterranean Sea. *Mar. Pollut. Bull., 138*, 260-265.
[http://dx.doi.org/10.1016/j.marpolbul.2018.11.008] [PMID: 30660272]

De-la-Torre, G. E., Dioses-Salinas, D. C., Pizarro-Ortega, C. I., Santillan, L. (2021). New plastic formations in the Anthropocene. *Sci. Total Environ., 754*, 142216. 28.
[http://dx.doi.org/10.1016/j.scitotenv.2020.142216]

Fossi, M.C., Pedà, C., Compa, M., Tsangaris, C., Alomar, C., Claro, F., Ioakeimidis, C., Galgani, F., Hema, T., Deudero, S., Romeo, T., Battaglia, P., Andaloro, F., Caliani, I., Casini, S., Panti, C., Baini, M. (2018). Bioindicators for monitoring marine litter ingestion and its impacts on Mediterranean biodiversity. *Environ. Pollut., 237*, 1023-1040.
[http://dx.doi.org/10.1016/j.envpol.2017.11.019] [PMID: 29153726]

Fortibuoni, T., Ronchi, F., Mačić, V., Mandić, M., Mazziotti, C., Peterlin, M., Prevenios, M., Prvan, M., Somarakis, S., Tutman, P., Bojanić Varezić, D., Virsek, M. K., Vlachogianni, T., Zeri, C. (2019). A harmonized and coordinated assessment of the abundance and composition of seafloor litter in the Adriatic-Ionian macroregion (Mediterranean Sea). *Mar. Pollut. Bull., 139*, 412e426.

Jambeck, J.R., Geyer, R., Wilcox, C., Siegler, T.R., Perryman, M., Andrady, A., Narayan, R., Law, K.L. (2015). Plastic waste inputs from land into the ocean. *Science, 347*(6223), 768-771.
[http://dx.doi.org/10.1126/science.1260352] [PMID: 25678662]

Li, W.C., Tse, H.F., Fok, L. (2016). Plastic waste in the marine environment: A review of sources, occurrence and effects. *Sci. Total Environ., 566-567*, 333-349.
[http://dx.doi.org/10.1016/j.scitotenv.2016.05.084] [PMID: 27232963]

Loulad, S., Houssa, R., Ouamari, N.E., Rhinane, H. (2019). Quantity and spatial distribution of seafloor marine debris in the Moroccan Mediterranean Sea. *Mar. Pollut. Bull., 139*, 163-173.
[http://dx.doi.org/10.1016/j.marpolbul.2018.12.036] [PMID: 30686415]

Maziane, F., Nachite, D., Anfuso, G. (2018). Artificial polymer materials debris characteristics along the moroccan mediterranean coast. *Mar. Pollut. Bull., 128*, 1-7.
[http://dx.doi.org/10.1016/j.marpolbul.2017.12.067] [PMID: 29571352]

Mghili, B., Analla, M., Aksissou, M. (2022). Face masks related to COVID-19 in the beaches of the Moroccan Mediterranean: An emerging source of plastic pollution. *Mar. Pollut. Bull., 174*, 113181.
[http://dx.doi.org/10.1016/j.marpolbul.2021.113181] [PMID: 34894579]

Mghili, B., Analla, M., Aksissou, M., Aissa, C. (2020). Marine debris in moroccan mediterranean beaches: an assessment of their abundance, composition and sources. *Mar. Pollut. Bull., 160*, 111692.
[http://dx.doi.org/10.1016/j.marpolbul.2020.111692] [PMID: 33181961]

Nachite, D., Meziane, F., Anfuso, G. (2018). Marine litter on Tangier urban beach (Morocco) sources and impacts for recreation. *Conference: Marine litter on Tangier urban beach (Morocco) sources and impacts for recreation.*Tetouan (Morocco).

Nachite, D., Maziane, F., Anfuso, G., Macias, A. Beach litter characteristics along the moroccan mediterranean coast: Implications for coastal zone management.*Beach Management Tools-Concepts, Methodologies and Case Studies.* (pp. 795-819). Cham: Springer.
[http://dx.doi.org/10.1007/978-3-319-58304-4_40]

Nachite, D., Maziane, F., Anfuso, G., Williams, A.T. (2019). Spatial and temporal variations of litter at the Mediterranean beaches of Morocco mainly due to beach users. *Ocean Coast. Manage., 179,* 104846.
[http://dx.doi.org/10.1016/j.ocecoaman.2019.104846]

Ocean Conservancy. (2011). *Tracking trash, 25 years of action for the ocean. In: 2011 Report.*

RGPH. (2014). (Recensement Général de la Population et de l'Habitat). Haut Comissariat au Plan, Morocco. Available from: http://rgphentableaux.hcp.ma/Default1/ (Accessed 28 july 2021.)

Suaria, G., Avio, C. G., Mineo, A., Lattin, G. L., Magaldi, M. G., Belmonte, G., Moore, C. J., Regoli, F., Aliani, S. (2016). The mediterranean plastic soup: Synthetic polymers in mediterranean surface waters. *Sci. Rep., 6,* 37551.
[http://dx.doi.org/10.1038/srep37551]

UNEP/MAPMEDPOL. (2011). Results of the assessment of the status of marine litter in the Mediterranean Sea. *In: UNEP/MAP (DEPI)/MED WG.357/Inf.4.*

Vlachogianni, T. (2019). Marine litter in mediterranean coastal and marine protected areas how bad is it? a snapshot assessment report on the amounts, composition and sources of marine litter found on beaches. *Interreg. Med.,* 1-40.

Vlachogianni, T. (2020). *Evaluation des déchets marins sur le littoral marocain.*

Zeri, C., Adamopoulou, A., Bojanić Varezić, D., Fortibuoni, T., Kovač Viršek, M., Kržan, A., Mandic, M., Mazziotti, C., Palatinus, A., Peterlin, M., Prvan, M., Ronchi, F., Siljic, J., Tutman, P., Vlachogianni, T. (2018). Floating plastics in Adriatic waters (Mediterranean Sea): From the macro- to the micro-scale. *Mar. Pollut. Bull., 136,* 341e350.

CHAPTER 5

Potential Ecological Impacts of Rare Earth Elements in the Marine Environment: A Baseline for Future Research

Ahmed Mandour[1,*]

[1] *Department of Oceanography, Faculty of Science, Alexandria University, Baghdad St., Moharem Bek, 21511, Alexandria, Egypt*

Abstract: Rare earth elements (REE) have become a strategic commodity of contemporary economies due to their various uses in the technological, smart, and renewable energy industries. The boom of their uses resulted in an increased influx to the marine environment either as a result of mining or industrial discharges, or from the disposal of solid wastes, atmospheric fallout for military tests of smart weapons, and remobilization from the sediments. Although sediments are the main reservoir of REE in the marine environment, and their auspicious normalization patterns are useful geochemical tracers, it has been found that anthropogenic contributions influence REE's natural occurrence. This indeed has raised concerns about the potential ecological impacts of REE on the marine biota and in turn on human health. The chapter gives some insights into the sources and potential ecological impacts of REE while revealing the need for future research and the knowledge gap about the REE and their ecological impacts as a group and as individual elements, as well as some potential solutions to the increased anthropogenic influx of REE to the marine environment. The potential ecological impacts of REE influx to the marine environment constitute both their bioavailability and their toxicity. Predicted ecological impacts on the marine biota may be similar to other trace metals, sharing analogous chemical characteristics. Nevertheless, whether LREE or HREE are more toxic is debatable, and their physiological and cytological effects on different organisms are still under investigation. This prompts the need for a new understanding of REE's ecological impacts by focusing on influx rates, ecotoxicity, and mitigation of ecological impacts.

Keywords: Aquatic, Anthropogenic contribution, Aquatic, Biota, Bioavailability, Bioconcentration, Ecological impacts, Ecotoxicity, Health, Lanthanides, Nano-oxides marine, Pollution, Pollution mitigation, Remobilization, Rare earth elements, Sediments.

---

[*] **Corresponding author Ahmed Mandour:** Department of Oceanography, Faculty of Science, Alexandria University, Baghdad St., Moharem Bek, 21511, Alexandria, Egypt; E-mail: ahmed.mandour@alexu.edu.eg

Tamer El-Sayed Ali (Ed.)

INTRODUCTION

The preservation of the marine environment is vital to the livelihood and well-being of humanity as a whole. The marine living and non-living resources and ecosystem services are invaluable to the current global economy and the sustainability of socio-economic development. As the need for economic expansion and technological development increased, the exerted pressures on the marine environment exponentially grew in response to a multitude of threats elicited by the interference between natural and man-made pressures.

One of the top concerns of such pressures is pollution by anthropogenic materials due to its ecological impacts as well as its impacts on human health and the economic viability of an ecosystem (UNEP, 2012). A substance can be considered a pollutant when it exceeds its natural background levels in a specific ecosystem either due to a deliberate or an accidental process. Trace metals are widely common inorganic pollutants, reported to be in contingency with industrial/agricultural hubs, densely populated urban centers, and areas with dense maritime activities (Forstner & Wittman, 1983; Berias, 2018). Consequently, trace metals belonging to the transitional elements such as V, Cr, Mn, Fe, Co, Ni, Cu, Zn, Cd, Pb, Hg, and As became the focus of research and were taken into consideration during designing pollution mitigation policies by states due to their common use in industry, agriculture, and domestic products.

In the past three decades, rare earth elements (REE) became widely used in technological industries. This prompted an increase in the demand for REE and increased the economic and strategic value of REE. During this period, limited attention was given to the ecological impacts of REE on the marine environment, with the aforementioned metals still the main focus of research and policymaking. In the 80s and the 90s, most research focused on the chemistry and geochemistry of REE in the marine environment. However, the exponential increase of reliability REE in modern industries which in turn resulted in an increase in their influx to the marine environment has started to grasp the attention of both researchers and policymakers as it holds potential ecological risks and threatens human health.

The present chapter probes the issue of increased REE influx to the marine environment as a form of pollution and explores the associated potential ecological risks while aiming also to shed the light on the relative mitigation options concerning mitigating those risks and points out the research gaps in such perspective.

RARE EARTH ELEMENTS AND THEIR SOURCES IN THE MARINE ENVIRONMENT

Rare earth elements (REE) are a chemically coherent group of metals. They comprise 17 elements including the 15 elements of lanthanides plus the other two elements (Scandium (Sc) & yttrium (Y)) that share similar chemical properties with Lanthanides (Lucas *et al.*, 2015). Lanthanides are classified according to their atomic weight into two groups (Fig. **1**): Light (LREE) includes the elements Lanthanum (La), cerium (Ce), praseodymium (Pr), neodymium (Nd) samarium (Sm), and europium(Eu) as well as scandium (Sc); while heavy (HREE) includes gadolinium (Gd), terbium (Tb), dysprosium (Dy), holmium (Ho), erbium (Er), thulium (Tm), ytterbium (Yb), and lutetium (Lu) (Gonzalez *et al.*, 2014). Some authors refer to Eu, Gd, Tb, and Dy as medium (MREE).

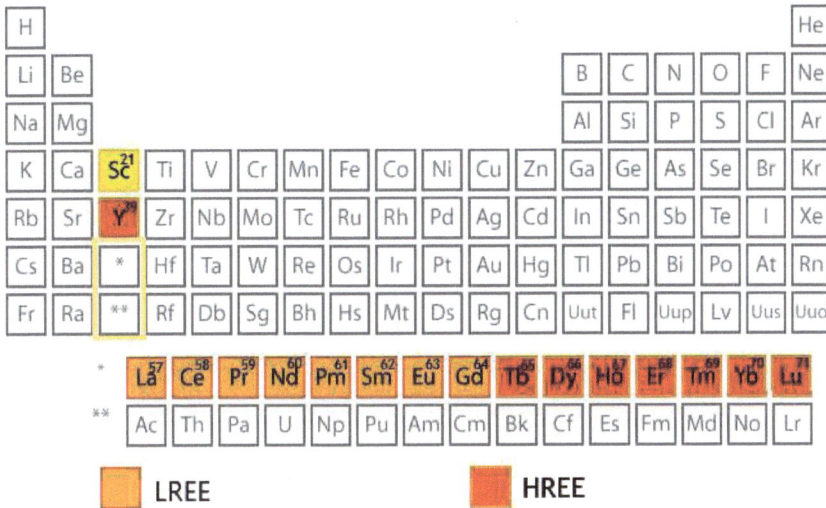

Fig. (1). The disposition of REE (LREE & HREE) in the periodic table.

REE occur naturally in the upper Earth's crust with Ce being the most abundant (60 ppm) ranking 25th of the 78 common elements in the Earth's crust millions, and Tm and Lu being the least abundant at about 0.5 part per million (Taylor & Mclennan, 1995). REE occurs either in economic ore deposits or in accessory minerals such as monazite. Economically viable REE deposits are abundant in Central Asia, specifically in China and Mongolia, however, the export restrictions imposed by China have drawn attention to other source countries such as Australia, Brazil, Canada, South Africa, Tanzania, Denmark (Greenland), and the United States (Fathollahzadeh *et al.*, 2019; Kennedy, 2016).

In the past decades, REE gained a huge economic value and became an essential component in cutting-edge technologies of communications and electronics, medical applications, and energy (Fig. **2**). The applications of REE include the manufacturing of chemical catalysts and electronics such as LED screens and smart chips. They are also used in the manufacture of lasers and optic fibers, the manufacture of high-strength magnets and alloys, and the manufacture of medical equipment especially medical imaging equipment, and in green technologies such as rechargeable batteries for electronics and hybrid/electric automobiles (Alzamly *et al.*, 2020; Balaram, 2019; Huang *et al.*, 2022; Lucas *et al.*, 2015a).

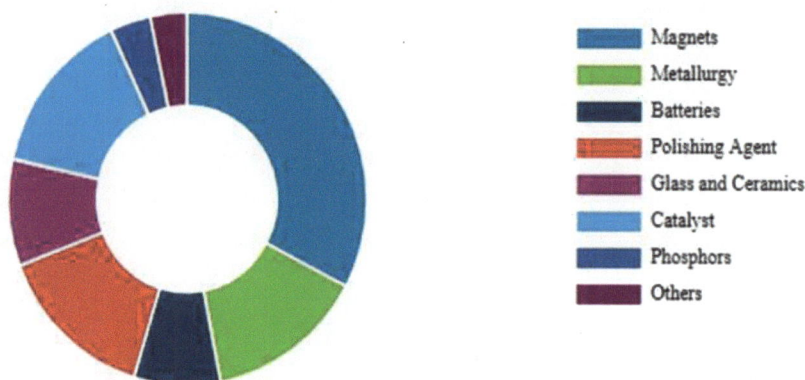

Fig. (2). Global REE market share in 2020 (data from www.fortunebusinessinsights.com).

Additionally, REE are widely used in advancing military technologies and the production of smart tactical weapons (Massari & Ruberti, 2013). Consequently, REE's strategic importance, akin to fossil fuels and other mineral resources, gained tremendous momentum and became part of international politics and geostrategic calculations of some nations such as China and the United States (Apergis & Apergis, 2017; Mancheri, 2016). For instance, REE were among the top EU imports and classified as strategic raw materials (Fig. **3**). In 2019, the Chinese government used REE exports to the US as a pressure card to dissuade the US from counter-policies.

Natural Sources of Ree in the Marine Environment

REE occurrence in the marine environment is attributed to both natural and anthropogenic sources (Di Leonardo *et al.*, 2009; Elderfield *et al.*, 1990a; Freslon *et al.*, 2014). Thus, to fully understand the role of anthropogenic contribution to the REE influx in the marine environment, the natural occurrence and controlling processes of REE in the marine environment as well as the relevant chemical and geochemical processes controlling their concentrations should be discerned.

Spain: Strontium	100%
China: Light rare earth elements	99%
China: Heavy rare earth elements	98%
China: Magnesium	93%
South Africa: Ruthenium*	93%
South Africa: Iridium*	92%
Turkey: Borates	89%
United States: Beryllium*	88%
Brazil: Niobium	85%
France: Hafnium	84%
South Africa: Rhodium*	80%
Chile: Lithium	78%

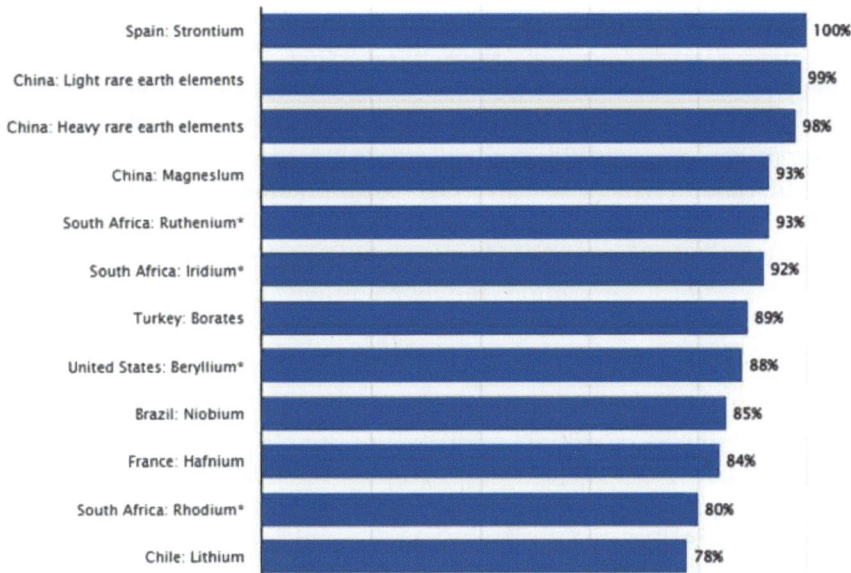

Fig. (3). Share of critical raw materials supplies in the EU in 2020, presented by main supplier nation and mineral (data from www.statista.com/).

Similar to other trace metals, the presence of REE in an aquatic environment takes one of three forms: (i) particulate REE which are associated with solid particles, (ii) dissolved REE which are associated with colloids and nanoparticles, and (iii) dissolved REE present in the water as ions or as chemical complexes (Dubinin, 2004; Elderfield *et al.*, 1990b).

The major holders of REE in the aquatic environment are the riverine, coastal, and marine sediments. REEs are either incorporated as a component of the minerals' matrices, adsorbed on the sediments' surface, or dissolved in the pore water. In the mineralogical sense, LREE are abundant in bastnasite and monazite, while HREE are usually present in apatite, cheralite, eudialyte, loparite, phosphorites, rare-earth-bearing clays (primarily through adsorption), secondary monazite, xenotime as well as spent uranium solutions (Dutta *et al.*, 2016). Although REE locked in mineral matrices are of limited ecological impacts, the ability of finer particles (silts and clays) to scavenge REE from water, in a mechanism probably similar in shape to other commonly studied trace metals, must play an essential role in the bioavailability of REE and increase their ecological impact under certain conditions. For instance, LREE is preferentially more adsorbed compared to HREE, except for silica phases. On the other hand, organic-rich materials are proven to have more affinity to scavenge REE from water (Byrne & Kim, 1990).

Geochemically normalized REE concentrations in marine sediments are a useful tracking signature that may characterize the sources of sediments and provide information about the current and previous sedimentary processes (Castañeda *et al.*, 2016; Mandour *et al.*, 2020). REE concentrations can be normalized to several background values to determine their enrichment/depletion patterns. The backgrounds include North American Shale Composite (NASC) (Gromet *et al.*, 1984), chondrite (Boynton, 1984), Australian Post-Archean Shale (PAAS) (McLennan *et al.*, 1993), and the Upper Continental Crust (UCC) (Taylor & McLennan, 1985). REE are good biogeochemical indicators (Chaudhuri & Cullers, 1979). They can be used as tracers for biogeochemical processes. They can be also used as precursors to water and sediment sources and fates. REE are associated with trace metals and thus can be used as a novel promising approach to anthropogenic impact evaluation (Tranchida *et al.*, 2011). REE are incorporated in marine sediments *via* several processes: (i) The composition of the source rocks is the primary controller (Mandour *et al.*, 2020), (ii) Scavenging from seawater (Elderfield *et al.*, 1988), (iii) Redox conditions during deposition and post-deposition (Abbott *et al.*, 2015), as well as anthropogenic influences (Piarulli *et al.*, 2021). In pore waters, three factors play a role in REE distribution and mobility: (i) anoxic conditions may induce the enrichment of MREE, (ii) particulate organic matter induces the enrichment of HREE, and (iii) Ce-oxide has a somehow anomalous pattern. This so-called "Ce-anomaly" does not exceed unity under any redox condition (Tostevin *et al.*, 2016).

Sources of dissolved REE that reach the marine waters include inputs from rivers and streams, hydrothermal activity, aeolian input, the diagenesis of sediments (Dubinin, 2004; Elderfield & Greaves, 1982; Leybourne & Johannesson, 2008). Rivers are major sources of dissolved REE (Elderfield *et al.*, 1990b; Ramesh *et al.*, 2000; Singh, 2009). Trivalent REE should exhibit strong, predominantly electrostatic complexes with ligands such as fluoride, sulfate, phosphate, carbonate, and hydroxide, however, they form weak complexes with chloride and nitrate, while they rarely attach to ammonia and bisulfide (Wood, 1990).

When released by hydrothermal vents, REE are associated with Fe-Mn oxides, particularly enriched in HREE, showing positive Ce-anomalies (enrichments) in the ferromanganese-oxyhydroxide phase and negative Y-anomalies in Ferro-oxyhydroxide phase (Surya Prakash *et al.*, 2012). However, hydrothermal Fe–Mn oxyhydroxides, which are incorporated into nodules in the hydrothermal-sedimentary deposits, show negative Ce anomalies as the REE composition of these deposits are developed during the adsorption of dissolved REE with a negative cerium anomaly and depletion in the LREE from deep waters (Dubinin, 2004).

Collectively, the natural presence of REE in the marine environment is attributed to inputs from rivers, local geochemical processes, and hydrothermal vents. Indeed, before attempting to investigate the anthropogenic inputs of REE in a given environment, the local background of pre-industrial concentrations of REE should be determined and the associated processes should be fairly understood. For example, the Nile deposits that reach the Mediterranean are rich in REE-bearing minerals and the observed anomalies are attributed to geochemical and weathering processes (Fielding *et al.*, 2017; Mandour *et al.*, 2020; Padoan *et al.*, 2011).

Anthropogenic Sources of Ree in the Marine Environment

The demand for REE in different technologies made them an important raw material in many industries. Whether deliberately or accidentally, this has significantly boosted their influx to the marine environment. REE can be released from multiple points and diffuse sources such as industrial facilities, medical and pharmaceutical facilities, mining and ore processing facilities, agricultural drainage, solid waste disposal, REE recycling plants, phosphatic fertilizers, and livestock feeds.

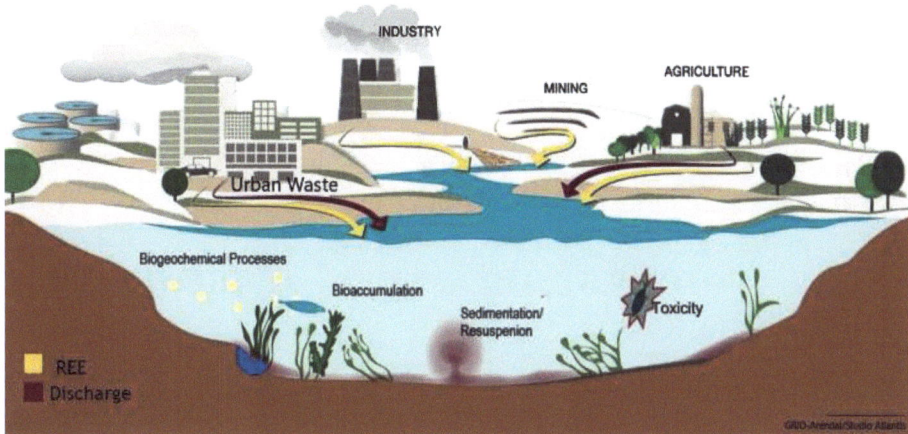

Fig. (4). Sources and fate of REE in the marine environment (modified on a template from www.grida.no/).

The first apparent source of anthropogenic REE in the marine environment is the mining of REE-containing deposits (El Zrelli *et al.*, 2021; Li *et al.*, 2013). The wastewater of acid mining reaches streams and the marine environment, significantly boosting REE concentrations in coastal waters. The treatment of REE-rich deposits incurs some occupational and environmental radiation exposure to naturally occurring radionuclides from the U and Th decay series as well as emanated radon. Variable contents of such radioactive materials should be assessed carefully during the selection of mining techniques (Barakos *et al.*, 2016).

Coastal areas that receive industrial discharges but have no mining activities have been observed to be contaminated by REE (Di Leonardo *et al.*, 2009). Additionally, high-tech wastes are contaminated by REE, and the unregulated discard of such wastes can result in them reaching drainage systems (Gwenzi *et al.*, 2018). Furthermore, urban wastewater treatment systems, which receive various industrial and municipal wastewater streams act as a source point of REE, particularly Gd (Lawrence *et al.*, 2006, 2009). Landfills act as REE reservoirs since they are often used for the disposal of post-consumer REE products such as televisions, computers, and cell phones. Seepage of water from those landfills can reach nearby aquatic environments and eventually to the coastal waters.

Agricultural activities have been observed to be a source of REE. Agricultural run-offs laden with agrochemicals were found to be contaminated by REE (Chaillou *et al.*, 2006; Xu *et al.*, 2002). Phosphate-based fertilizers are produced from phosphate ores that contain REE at various concentrations (El Zrelli *et al.*, 2021). For example, it was estimated that around 50-100 million tons of REE oxides enter the Chinese agricultural systems (Liang *et al.*, 2005). Although the use of REE is not a requirement in agricultural activity, the sheer amount of the discharged wastewater and leached sediments from agriculture might introduce significant amounts of REE to the marine environment. Leaching REE and other metals from agricultural soil can be considered a collateral type of pollution. The recipient aquatic environment of agricultural discharge will be enriched with REE and reach contamination levels.

POTENTIAL ECOLOGICAL IMPACTS OF REE IN THE MARINE ENVIRONMENT

The bioavailability of all metals including REE in the marine ecosystem is influenced by abiotic factors such as pH, concentration and type of organic and inorganic ligands, and temperature (Gonzalez *et al.*, 2014). The solubility of REE plays a key role in their mobility and thus their bioavailability as well as it influences competitive interactions with inorganic ions of hydrogen, calcium, and magnesium (Sholkovitz, 1992). The potentiality of an ecological impact resulting from REE on aquatic biota is dependent on how much REE can be accumulated in the cells or tissues of different types of organisms and specific species. Most of the bioaccumulation studies focused on LREE thus the full ecological impact of the REE on aquatic biota is still understudied.

REE are generally characterized by low mobility, consequently, it is a good tracer of sedimentary and pedogenic processes (Laveuf *et al.*, 2008; Padoan *et al.*, 2015; Singh, 2010). Eu and Ce are used as indicators of geochemical weathering (Mandour *et al.*, 2020). On the other hand, the low mobility of REE has raised the risk of being bioaccumulated in organisms in contact with sediments (Xinde *et al.*, 2000).

There is some contrast among different studies showing a contradiction to the general concept that "the higher the atomic number; the higher the toxicity". The bioconcentration factors (BCF), which is the ratio between the elemental concentration in an organism and the surrounding media, in freshwater organisms such as *Lemna minor* and *Potamogeton pectinatus*, have shown that LREE concentrations in living organisms and water are quite similar. Inversely, the BCF of HREE is lesser in organisms than in the surrounding pore waters. This study attributed this phenomenon to the increase in ligand stability constants with increasing the atomic number that preferentially kept the heavier lanthanides in solution bound to available ligands (Weltje *et al.*, 2002). In such cases, the typical 'saw-tooth' lanthanide pattern implies that lanthanides are transported as a coherent group through aquatic ecosystems, with observed anomalies in Ce and Eu and could be explained by their redox chemistry. Shells also contained much lower concentrations and were relatively enriched in Eu, and to a lesser extent in Ce. In different species of gram-positive bacteria, opposing results showed that BCF is directly proportional to increasing the atomic number of REE, however, the results were species-dependent (Tsuruta, 2006). The contrasting results between the two studies indicate that different species can accumulate either HREE or LREE and that bioavailability differs according to both the concentration of REE in the media as well as the tendency of the organism to accumulate different elements.

The ecotoxicity study of Ce (LREE), Gd (MREE), and Lu (HREE) in Crustaceans (Branchiopoda and Ostracoda), Cnidaria, Rotifera, Chlorophyta, and proteobacteria revealed that different marine organisms display different responses to exposure to REE (González *et al.*, 2015). Chlorophyta and Proteobacteria were more sensitive to HREE, while other organisms were equally sensitive to Ce, Gd, and Lu. Another study on Mediterranean Seaweeds revealed that areas contaminated with trace metals are also contaminated with REE. The seaweeds showed high contamination factors for Ce, Yb, Pr, La, and Nd (Squadrone *et al.*, 2017).

The widespread use of REE as nano-oxides triggered a lot of research done on the toxicity of lanthanides on aquatic organisms (Blinova *et al.*, 2020). La and Ce are the most commonly studied REE. Toxicity tests are used to assess the risk of lanthanide-based nanoparticles, however, these tests have limited relevance to the actual ecotoxicity of REE in the aquatic environment (Merrington *et al.*, 2014). In such a case, REE behavior in the natural environment is quite different from *in vitro* conditions.

There are two detrimental factors of REE's potential ecological impacts. First the amount of influx rate to a certain area. Human activities that result in REE-laden wastewaters or REE-laden sediments will determine the degree of contamination in this area. REE from different activities can mix up with other organic and inorganic pollutants. This mixing can have a cascade effect that amplifies the amount of REE that reaches the marine environment. For instance, particulate organic pollutants (POPs) and sediments can scavenge more metals and REEs from inland drainage to the marine environment and increase their concentrations in waters and sediments. The second factor is the bioavailability of REE, which is dependent on their atomic weight and the chemical and physical conditions of the ecosystem. However, the toxic response of the living organisms in a contaminated area is species-dependent and a lot is still there to be understood.

Medical studies have shown that REE have strong adverse effects on human health (Li *et al.*, 2013; Pagano *et al.*, 2019). The contamination of the marine environment by REE will inherently affect humans through the consumption of REE-contaminated seafood. Similar to trace metals, edible organisms that bioaccumulate REE to high concentrations will have toxic effects on human consumers (Jiang *et al.*, 2012; Kapranov *et al.*, 2021). It is very important to consider that the increased concentration of REE in seafood consumed by humans may result in unseen impacts. For example, physicians strongly advise against High-Hg fish for pregnant women. This should make researchers similarly think about REE, particularly with their high atomic numbers and similarities to other trace metals.

RESEARCH GAPS AND MITIGATION PERSPECTIVES

The present chapter summarized the general traits of research about the anthropogenic influx and the ecological impacts of REE in the marine environment. Despite the availability of literature about REE, there is a considerable knowledge gap about its potential ecological impacts in general, and in the marine environment, in specific. The standing knowledge situation guides the focus of research towards four points.

The first focus is on understanding the influx and fate of anthropogenic REE in the marine environment. Since the 1980s, a considerable amount of research and data has been produced on the marine chemistry and geochemistry of REE, however, there is a need to apprehend the amount of REE influx from different sources, their pathways to the marine environment, and their fate. In such a sense, the biogeochemical cycle of REE should be understood on both regional and global levels.

The second focus is on estimating and evaluating the degree of ecological impacts of REE. Two approaches can be employed to determine such ecological risk, either in the laboratory through using spiked solutions of single elements or mixtures of elements on selected aquatic organisms, or in the natural environment through measuring the concentrations of REE and their observable impacts on the organisms. The first approach is quite useful to determine numerically the lethal doses and the degree of toxicity, and the toxic effects on the organisms can be studied firsthand. However, REE's behavioral differences in the experimental and the natural environment hamper the dependability of such results. Furthermore, the concentrations used in the lab approach are sometimes much higher than the possible concentrations in the natural environment, thus, the findings are limited to determining chemical toxicity rather than ecotoxicity.

On the other hand, the second approach can assess the potential ecological impacts of REE in the natural environment by measuring the concentrations of REE in sediments, pore waters, water columns, and living organisms. The obtained data can be linked directly to potential ecological impacts in the form of bioavailability, bioconcentration, elemental mobility, ecotoxicity, biomass productivity, and biological deformation (morphological, physiological, or genetic). However, the areas under investigation are usually polluted by other trace metals which can interfere with the findings of the studies and increase the ecological impacts on living organisms. Additionally, REE has very low concentrations in the water column which can be masked by other elements, which makes building assumptions very difficult for researchers. Consequently, the ecological risk factors for low REE concentrations must be established by combining both approaches and discerning the ecological impacts of REE from other trace metals. Additionally, establishing a unified threshold effect concentration of REE and the degree of its toxicity within the natural environment is key for comprehensive ecological risk evaluation studies.

One of the challenges to evaluating ecological risk is harmonizing the data and the backgrounds reported in the studies to allow comparability and form a cohesive point of view. Normalized patterns are recommended to distinguish peaks and anomalies in water and sediments. These patterns should be reflected in biological studies as the environmental background values (Piarulli *et al.*, 2021).

Although the theme of this chapter is about the marine environment, however, if the ecological risks of REE are to be thoroughly studied, their impacts on human health should also be taken into consideration. This includes the estimated amount of REE that are be consumed by humans through different sources and their adverse effects on pregnancy, nursing, children, and adults.

The third focus should be the individualization of elements of the REE group. Most studies treat REE as a homogenous group, either dealt with collectively or represented by elements of LREE, MREE, and HREE. This is suitable in the case of geochemical studies because REE are known to be chemically homogenous and have analogous behavior. However, in the case of contamination studies, individual elements have different sources, and those point and diffuse sources will subsequently have different influx rates, bioavailability, and might be even behavioral anomalies. Individual influx, fate, and behavior of elements should be studied as a pretext to studying their toxicological impacts on the organism of a certain contaminated area.

The fourth and the most important focus should be the mitigation measures needed to reduce contamination by REE in the natural environment. This has two aspects. The first is the ecological impacts of REE and the consequent impacts on human health can be reduced by improving industrial and agricultural wastewaters treatment technologies, improving the recycling technologies of REE-laden solid wastes such as old electronic devices and batteries, and promoting policies that prevent the disposal of solid and liquid wastes from high-tech and medical industries. The second aspect is based on the fact that REE have such high economic value and a lot of industrial uses that the anthropogenic influx of REE to the marine environment should be considered a waste of valuable resources. Thus, mitigation policies and removal technologies should be centered around the recycling and reuse of REE as much as the conservation of the marine environment.

Several strategies are suggested to reduce the influx of REE to the marine environment: (1) There is a need for the development of sustainable technologies for raw ore extraction to face the scarcity of REE, and (2) Alternatives for REE in the industry can reduce the demand and waste either by material substitution such as carbon nanomaterials (CNMs) or by technology replacement (Gwenzi *et al.*, 2018). Direct recycling of REE-laden wastes is still an infant approach. There are multiple challenges to reusing and recycling high-tech wastes and they are still largely limited to laboratory-scale studies (Tan *et al.*, 2015). Algae-based technology is considered one of the most promising methods for recovering REE for its high efficiency, low cost, and wide applicability (Cao *et al.*, 2021).

REE products such as small and personal electronic devices are so widely dispersed and thus very challenging to recycle, however, policy adjustments can help in this case. For example, incentives and price cuts on new devices in an exchange for old or replaced devices will encourage the public to hand over theirs and prevent their regulated disposal plus allow the recycling of its materials to prevent wasting REE resources. State-of-the-art technologies for the capture and recovery of REE from wastewater are thus essential for a sustainable future.

CONCLUSION

It is evident from the current research that the multitudinous uses of REE in high-tech industries steered concerns about their influx to the marine environment *via* multiple sources. Nevertheless, these concerns are yet to materialize into focusing on the knowledge gap about biogeochemistry and the ecotoxicity of REE. Designing and implementing policies for the sustainable extraction, use, and disposal of REE and REE-containing products are necessary to mitigate their predicted ecological impacts. The strategic importance of REE in high-tech and military industries may mask the effort to diminish their ecological impact, particularly due to their role in manufacturing renewable energy technology. The reduction of REE's influx to the natural environment is intertwined with the need to leverage the conservation of REE reserves (for ore-producing countries) and reduce dependency on imports (for non-producing countries). The economic value and the scarcity of REE highlight the need for recycling discarded devices and systems containing REE, which will have a on the conservation of resources and pollution prevention.

REFERENCES

Abbott, A.N., Haley, B.A., McManus, J., Reimers, C.E. (2015). The sedimentary flux of dissolved rare earth elements to the ocean. *Geochim. Cosmochim. Acta, 154*, 186-200.
[http://dx.doi.org/10.1016/j.gca.2015.01.010]

Alzamly, A., Bakiro, M., Hussein Ahmed, S., Alnaqbi, M.A., Nguyen, H.L. (2020). Rare-earth metal–organic frameworks as advanced catalytic platforms for organic synthesis. *Coord. Chem. Rev., 425*, 213543.
[http://dx.doi.org/10.1016/j.ccr.2020.213543]

Apergis, E., Apergis, N. (2017). The role of rare earth prices in renewable energy consumption: The actual driver for a renewable energy world. *Energy Econ., 62*, 33-42.
[http://dx.doi.org/10.1016/j.eneco.2016.12.015]

Barakos, G., Mischo, H., Gutzmer, J. (2016). Rare earth underground mining approaches with respect to radioactivity control and monitoring strategies. *Rare Earths Ind. Technol. Econ. Environ. Implic.,* (Jan), 121-138.
[http://dx.doi.org/10.1016/B978-0-12-802328-0.00008-5]

Beiras, R. (2018). Part I: Pollutants in marine ecosystems. *in Marine Pollution: Sources, Fate and Effects of Pollutants in Coastal Ecosystems,* 1-157.

Balaram, V. (2019). Rare earth elements: A review of applications, occurrence, exploration, analysis, recycling, and environmental impact. *Geoscience Frontiers, 10*(4), 1285-1303.

[http://dx.doi.org/10.1016/j.gsf.2018.12.005]

Blinova, I., Muna, M., Heinlaan, M., Lukjanova, A., Kahru, A. (2020). Potential hazard of lanthanides and lanthanide-based nanoparticles to aquatic ecosystems: Data gaps, challenges and future research needs derived from bibliometric analysis. *Nanomaterials, 10*(2), 328.
[http://dx.doi.org/10.3390/nano10020328] [PMID: 32075069]

Byrne, R.H., Kim, K.H. (1990). Rare earth element scavenging in seawater. *Geochim. Cosmochim. Acta, 54*(10), 2645-2656.
[http://dx.doi.org/10.1016/0016-7037(90)90002-3]

Boynton, W.V. (1984). Cosmochemistry of the rare earth elements: Meteorite studies. *Develop. Geochem., 2*, 63-114.
[http://dx.doi.org/10.1016/B978-0-444-42148-7.50008-3]

Cao, Y., Shao, P., Chen, Y., Zhou, X., Yang, L., Shi, H., Yu, K., Luo, X., Luo, X. (2021). A critical review of the recovery of rare earth elements from wastewater by algae for resources recycling technologies. *Resour. Conserv. Recycling, 169*, 105519.
[http://dx.doi.org/10.1016/j.resconrec.2021.105519]

Castañeda, I.S., Schouten, S., Pätzold, J., Lucassen, F., Kasemann, S., Kuhlmann, H., Schefuß, E. (2016). Hydroclimate variability in the Nile River Basin during the past 28,000 years. *Earth Planet. Sci. Lett., 438*, 47-56.
[http://dx.doi.org/10.1016/j.epsl.2015.12.014]

Chaudhuri, S., Cullers, R.L. (1979). The distribution of rare-earth elements in deeply buried Gulf Coast sediments. *Chem. Geol., 24*(3-4), 327-338.
[http://dx.doi.org/10.1016/0009-2541(79)90131-1]

Chaillou, G., Anschutz, P., Lavaux, G., Blanc, G. (2006). Rare earth elements in the modern sediments of the Bay of Biscay (France). *Mar. Chem., 100*(1-2), 39-52.
[http://dx.doi.org/10.1016/j.marchem.2005.09.007]

Di Leonardo, R., Bellanca, A., Neri, R., Tranchida, G., Mazzola, S. (2009). Distribution of REEs in box-core sediments offshore an industrial area in SE Sicily, Ionian Sea: Evidence of anomalous sedimentary inputs. *Chemosphere, 77*(6), 778-784.
[http://dx.doi.org/10.1016/j.chemosphere.2009.08.021] [PMID: 19735932]

Dubinin, A.V. (2004). Geochemistry of rare earth elements in the ocean. *Lithol. Miner. Resour., 39*(4), 289-307.
[http://dx.doi.org/10.1023/B:LIMI.0000033816.14825.a2]

Dutta, T., Kim, K.H., Uchimiya, M., Kwon, E.E., Jeon, B.H., Deep, A., Yun, S.T. (2016). Global demand for rare earth resources and strategies for green mining. *Environ. Res., 150*, 182-190.
[http://dx.doi.org/10.1016/j.envres.2016.05.052] [PMID: 27295408]

UNEP. (2012). 21 issues for the 21st century : Results of the UNEP foresight process on emerging environmental issues.
[http://dx.doi.org/10.1016/j.envdev.2012.03.005]

Forstner, U., Wittmann, G. T. (1983). *Metals Pollution in the Aquatic Environment.* Springer Berlin, Heidelberg.

Taylor, S.R., McLennan, S.M. (1985). The Continental Crust: Its Composition and Evolution. Blackwell Scientific Publications.
[http://dx.doi.org/10.1002/gj.3350210116]

El Zrelli, R., Baliteau, J.Y., Yacoubi, L., Castet, S., Grégoire, M., Fabre, S., Sarazin, V., Daconceicao, L., Courjault-Radé, P., Rabaoui, L. (2021). Rare earth elements characterization associated to the phosphate fertilizer plants of Gabes (Tunisia, Central Mediterranean Sea): Geochemical properties and behavior, related economic losses, and potential hazards. *Sci. Total Environ., 791*, 148268.
[http://dx.doi.org/10.1016/j.scitotenv.2021.148268] [PMID: 34139493]

Elderfield, H., Upstill-Goddard, R., Sholkovitz, E.R. (1990). The rare earth elements in rivers, estuaries, and coastal seas and their significance to the composition of ocean waters. *Geochim. Cosmochim. Acta, 54*(4), 971-991.
[http://dx.doi.org/10.1016/0016-7037(90)90432-K]

Elderfield, H., Upstill-Goddard, R., Sholkovitz, E.R. (1990). The rare earth elements in rivers, estuaries, and coastal seas and their significance to the composition of ocean waters. *Geochim. Cosmochim. Acta, 54*(4), 971-991.
[http://dx.doi.org/10.1016/0016-7037(90)90432-K]

Elderfield, H., Whitfield, M., Burton, J.D., Bacon, M.P., Liss, P.S. (1988). The oceanic chemistry of the rare-earth elements. *Philos. Trans. R. Soc. Lond. A, 325*(1583), 105-126.
[http://dx.doi.org/10.1098/rsta.1988.0046]

Elderfield, H., Greaves, M.J. (1982). The rare earth elements in seawater. *Nature, 296*(5854), 214-219.
[http://dx.doi.org/10.1038/296214a0]

Fathollahzadeh, H., Khaleque, H.N., Eksteen, J., Kaksonen, A.H., Watkin, E.L.J. (2019). Effect of glycine on bioleaching of rare earth elements from Western Australian monazite by heterotrophic and autotrophic microorganisms. *Hydrometallurgy, 189*, 105137.
[http://dx.doi.org/10.1016/j.hydromet.2019.105137]

Fielding, L., Najman, Y., Millar, I., Butterworth, P., Ando, S., Padoan, M., Barfod, D., Kneller, B. (2017). A detrital record of the Nile River and its catchment. *J. Geol. Soc., 174*(2), 301-317.
[http://dx.doi.org/10.1144/jgs2016-075]

Freslon, N., Bayon, G., Toucanne, S., Bermell, S., Bollinger, C., Chéron, S., Etoubleau, J., Germain, Y., Khripounoff, A., Ponzevera, E., Rouget, M-L. (2014). Rare earth elements and neodymium isotopes in sedimentary organic matter. *Geochim. Cosmochim. Acta, 140*(September), 177-198.
[http://dx.doi.org/10.1016/j.gca.2014.05.016]

González, V., Vignati, D.A.L., Pons, M.N., Montarges-Pelletier, E., Bojic, C., Giamberini, L. (2015). Lanthanide ecotoxicity: First attempt to measure environmental risk for aquatic organisms. *Environ. Pollut., 199*, 139-147.
[http://dx.doi.org/10.1016/j.envpol.2015.01.020] [PMID: 25645063]

Gonzalez, V., Vignati, D.A.L., Leyval, C., Giamberini, L. (2014). Environmental fate and ecotoxicity of lanthanides: Are they a uniform group beyond chemistry? *Environ. Int., 71*, 148-157.
[http://dx.doi.org/10.1016/j.envint.2014.06.019] [PMID: 25036616]

Gromet, L.P., Haskin, L.A., Korotev, R.L., Dymek, R.F. (1984). The *North American shale composite* : Its compilation, major and trace element characteristics. *Geochim. Cosmochim. Acta, 48*(12), 2469-2482.
[http://dx.doi.org/10.1016/0016-7037(84)90298-9]

Gwenzi, W., Mangori, L., Danha, C., Chaukura, N., Dunjana, N., Sanganyado, E. (2018). Sources, behaviour, and environmental and human health risks of high-technology rare earth elements as emerging contaminants. *Sci. Total Environ., 636*, 299-313.
[http://dx.doi.org/10.1016/j.scitotenv.2018.04.235] [PMID: 29709849]

Huang, Y., Zhai, X., Ma, T., Zhang, M., Pan, H., Weijia Lu, W., Zhao, X., Sun, T., Li, Y., Shen, J., Yan, C., Du, Y. (2022). Rare earth-based materials for bone regeneration: Breakthroughs and advantages. *Coord. Chem. Rev., 450*, 214236.
[http://dx.doi.org/10.1016/j.ccr.2021.214236]

Jiang, D.G., Yang, J., Zhang, S., Yang, D.J. (2012). A survey of 16 rare Earth elements in the major foods in China. *Biomed. Environ. Sci., 25*(3), 267-271.
[http://dx.doi.org/10.3967/0895-3988.2012.03.003] [PMID: 22840576]

Kapranov, S.V., Karavantseva, N.V., Bobko, N.I., Ryabushko, V.I., Kapranova, L.L. (2021). Element contents in three commercially important edible mollusks harvested off the southwestern coast of crimea (Black sea) and assessment of human health risks from their consumption. *Foods, 10*(10), 2313.

[http://dx.doi.org/10.3390/foods10102313] [PMID: 34681363]

Kennedy, J.C. (2016). Rare earth production, regulatory usa/international constraints and chinese dominance. *Rare Earths Ind. Technol. Econ. Environ. Implic.,* 37-55.
[http://dx.doi.org/10.1016/B978-0-12-802328-0.00003-6]

Lawrence, M.G., Ort, C., Keller, J. (2009). Detection of anthropogenic gadolinium in treated wastewater in South East Queensland, Australia. *Water Res.,* 43(14), 3534-3540.
[http://dx.doi.org/10.1016/j.watres.2009.04.033] [PMID: 19541341]

Lawrence, M.G., Greig, A., Collerson, K.D., Kamber, B.S. (2006). Rare earth element and yttrium variability in South East Queensland waterways. *Aquat. Geochem.,* 12(1), 39-72.
[http://dx.doi.org/10.1007/s10498-005-4471-8]

Laveuf, C., Cornu, S., Juillot, F. (2008). Rare earth elements as tracers of pedogenetic processes. *C. R. Geosci.,* 340(8), 523-532.
[http://dx.doi.org/10.1016/j.crte.2008.07.001]

Leybourne, M.I., Johannesson, K.H. (2008). Rare earth elements (REE) and yttrium in stream waters, stream sediments, and Fe–Mn oxyhydroxides: Fractionation, speciation, and controls over REE+Y patterns in the surface environment. *Geochim. Cosmochim. Acta,* 72(24), 5962-5983.
[http://dx.doi.org/10.1016/j.gca.2008.09.022]

Li, X., Chen, Z., Chen, Z., Zhang, Y. (2013). A human health risk assessment of rare earth elements in soil and vegetables from a mining area in Fujian Province, Southeast China. *Chemosphere,* 93(6), 1240-1246.
[http://dx.doi.org/10.1016/j.chemosphere.2013.06.085] [PMID: 23891580]

Liang, T., Zhang, S., Wang, L., Kung, H.T., Wang, Y., Hu, A., Ding, S. (2005). Environmental biogeochemical behaviors of rare earth elements in soil–plant systems. *Environ. Geochem. Health,* 27(4), 301-311.
[http://dx.doi.org/10.1007/s10653-004-5734-9] [PMID: 16027965]

Lucas, J., Lucas, P., Le Mercier, T., Rollat, A., Davenport, W. (2015). Rare earth electronic structures and trends in properties. *Rare Earths,* (Jan), 123-139.
[http://dx.doi.org/10.1016/B978-0-444-62735-3.00008-5]

Lucas, J., Lucas, P., Le Mercier, T., Rollat, A., Davenport, W. (2015). Overview. *Rare Earths,* 1826(Jan), 1-14.
[http://dx.doi.org/10.1016/B978-0-444-62735-3.00001-2] [PMID: 30194590]

Massari, S., Ruberti, M. (2013). Rare earth elements as critical raw materials: Focus on international markets and future strategies. *Resour. Policy,* 38(1), 36-43.
[http://dx.doi.org/10.1016/j.resourpol.2012.07.001]

Mancheri, N.A. (2016). An overview of chinese rare earth export restrictions and implications. *Rare Earths Ind. Technol. Econ. Environ. Implic.,* 21-36.
[http://dx.doi.org/10.1016/B978-0-12-802328-0.00002-4]

Mandour, A.S., Ghezzi, L., Lezzerini, M., El-Gamal, A.A., Petrini, R., Elshazly, A. (2020). Geochemical characterization of recent Nile Delta inner shelf sediments: Tracing natural and human-induced alterations into a deltaic system. *Egypt. J. Aquat. Res.,* 46(4), 355-361.
[http://dx.doi.org/10.1016/j.ejar.2020.10.002]

McLennan, S.M., Hemming, S., McDaniel, D.K., Hanson, G.N. (1993). Geochemical approaches to sedimentation, provenance, and tectonics.*Proces. Control. Comp. Clas. Sedim.* Geological Society of America.
[http://dx.doi.org/10.1130/SPE284-p21]

Merrington, G., An, Y.J., Grist, E.P.M., Jeong, S.W., Rattikansukha, C., Roe, S., Schneider, U., Sthiannopkao, S., Suter, G.W., II, Van Dam, R., Van Sprang, P., Wang, J.Y., Warne, M.S.J., Yillia, P.T., Zhang, X.W., Leung, K.M.Y. (2014). Water quality guidelines for chemicals: Learning lessons to deliver meaningful environmental metrics. *Environ. Sci. Pollut. Res. Int.,* 21(1), 6-16.

[http://dx.doi.org/10.1007/s11356-013-1732-8] [PMID: 23619928]

Padoan, M., Garzanti, E., Harlavan, Y., Villa, I.M. (2011). Tracing Nile sediment sources by Sr and Nd isotope signatures (Uganda, Ethiopia, Sudan). *Geochim. Cosmochim. Acta, 75*(12), 3627-3644. [http://dx.doi.org/10.1016/j.gca.2011.03.042]

Padoan, M., Vezzoli, G., Garzanti, E., El Kammar, A. (2015). The modern Nile sediment system : Processes and products *Quat. Sci. Rev., 130*(15), 9-56. [http://dx.doi.org/10.1016/j.quascirev.2015.07.011]

Pagano, G., Thomas, P.J., Di Nunzio, A., Trifuoggi, M. (2019). Human exposures to rare earth elements: Present knowledge and research prospects. *Environ. Res., 171*, 493-500. [http://dx.doi.org/10.1016/j.envres.2019.02.004] [PMID: 30743241]

Piarulli, S., Hansen, B.H., Ciesielski, T., Zocher, A.L., Malzahn, A., Olsvik, P.A., Sonne, C., Nordtug, T., Jenssen, B.M., Booth, A.M., Farkas, J. (2021). Sources, distribution and effects of rare earth elements in the marine environment: Current knowledge and research gaps. *Environ. Pollut., 291*, 118230. [http://dx.doi.org/10.1016/j.envpol.2021.118230] [PMID: 34597732]

Ramesh, R., Ramanathan, A., Ramesh, S., Purvaja, R., Subramanian, V. (2000). Distribution of rare earth elements and heavy metals in the surficial sediments of the Himalayan river system. *Geochem. J., 34*(4), 295-319. [http://dx.doi.org/10.2343/geochemj.34.295]

Sholkovitz, E.R. (1992). Chemical evolution of rare earth elements: fractionation between colloidal and solution phases of filtered river water. *Earth Planet. Sci. Lett., 114*(1), 77-84. [http://dx.doi.org/10.1016/0012-821X(92)90152-L]

Singh, P. (2010). Geochemistry and provenance of stream sediments of the Ganga River and its major tributaries in the Himalayan region, India. *Chem. Geol., 269*(3-4), 220-236. [http://dx.doi.org/10.1016/j.chemgeo.2009.09.020]

Singh, P. (2009). Major, trace and REE geochemistry of the Ganga River sediments: Influence of provenance and sedimentary processes. *Chem. Geol., 266*(3-4), 242-255. [http://dx.doi.org/10.1016/j.chemgeo.2009.06.013]

Surya Prakash, L., Ray, D., Paropkari, A.L., Mudholkar, A.V., Satyanarayanan, M., Sreenivas, B., Chandrasekharam, D., Kota, D., Kamesh Raju, K.A., Kaisary, S., Balaram, V., Gurav, T. (2012). Distribution of REEs and yttrium among major geochemical phases of marine Fe–Mn-oxides: Comparative study between hydrogenous and hydrothermal deposits. *Chem. Geol., 312-313*, 127-137. [http://dx.doi.org/10.1016/j.chemgeo.2012.03.024]

Squadrone, S., Brizio, P., Battuello, M., Nurra, N., Sartor, R.M., Benedetto, A., Pessani, D., Abete, M.C. (2017). A first report of rare earth elements in northwestern Mediterranean seaweeds. *Mar. Pollut. Bull., 122*(1-2), 236-242. [http://dx.doi.org/10.1016/j.marpolbul.2017.06.048] [PMID: 28647152]

Tan, Q., Li, J., Zeng, X. (2015). Rare earth elements recovery from waste fluorescent lamps: A review. *Crit. Rev. Environ. Sci. Technol., 45*(7), 749-776. [http://dx.doi.org/10.1080/10643389.2014.900240]

Tranchida, G., Oliveri, E., Angelone, M., Bellanca, A., Censi, P., D'Elia, M., Neri, R., Placenti, F., Sprovieri, M., Mazzola, S. (2011). Distribution of rare earth elements in marine sediments from the Strait of Sicily (western Mediterranean Sea): Evidence of phosphogypsum waste contamination. *Mar. Pollut. Bull., 62*(1), 182-191. [http://dx.doi.org/10.1016/j.marpolbul.2010.11.003] [PMID: 21130477]

Tsuruta, T. (2006). Selective accumulation of light or heavy rare earth elements using gram-positive bacteria. *Colloids Surf. B Biointerfaces, 52*(2), 117-122. [http://dx.doi.org/10.1016/j.colsurfb.2006.04.014] [PMID: 16797944]

Weltje, L., Heidenreich, H., Zhu, W., Wolterbeek, H.T., Korhammer, S., de Goeij, J.J.M., Markert, B. (2002).

Lanthanide concentrations in freshwater plants and molluscs, related to those in surface water, pore water and sediment. A case study in The Netherlands. *Sci. Total Environ., 286*(1-3), 191-214.
[http://dx.doi.org/10.1016/S0048-9697(01)00978-0] [PMID: 11887873]

Wood, S.A. (1990). The aqueous geochemistry of the rare-earth elements and yttrium. *Chem. Geol., 82*(C), 159-186.
[http://dx.doi.org/10.1016/0009-2541(90)90080-Q]

Xinde, C., Xiaorong, W., Guiwen, Z. (2000). Assessment of the bioavailability of rare earth elements in soils by chemical fractionation and multiple regression analysis. *Chemosphere, 40*(1), 23-28.
[http://dx.doi.org/10.1016/S0045-6535(99)00225-8] [PMID: 10665441]

Xu, X., Zhu, W., Wang, Z., Witkamp, G.J. (2002). Distributions of rare earths and heavy metals in field-grown maize after application of rare earth-containing fertilizer. *Sci. Total Environ., 293*(1-3), 97-105.
[http://dx.doi.org/10.1016/S0048-9697(01)01150-0] [PMID: 12109484]

SUBJECT INDEX

www.ingramcontent.com/pod-product-compliance
Lightning Source LLC
Chambersburg PA
CBHW041447210326
41599CB00004B/158